Osteoimmunology

For other titles published in this series, go to
www.springer.com/series/5584

Recent Volumes in this Series

Volume 651
DEVELOPMENT AND ENGINEERING OF DOPAMINE NEURONS
Edited by R. Jeroen Pasterkamp, Marten P. Smidt, and J. Peter H. Burbach

Volume 652
INHERITED NEUROMUSCULAR DISEASES: TRANSLATION FROM PATHOMECHANISMS TO THERAPIES
Edited by Carmen Espinós, Vincente Felipo, and Francesc Palau

Volume 653
TARGET PATTERN RECOGNITION IN INNATE IMMUNITY
Edited by Uday Kishore

Volume 654
THE ISLETS OF LANGERHANS
Edited by Md. Shahidul Islam

Volume 655
PHARMACEUTICAL BIOTECHNOLOGY
Edited by Carlos A. Guzmán and Giora Z. Feuerstein

Volume 656
APC PROTEINS
Edited by Inke S. Näthke and Brooke M. McCarthney

Volume 657
BRAIN INSPIRED COGNITIVE SYSTEMS 2008
Edited by Amir Hussain, Igor Aleksander, Leslie S. Smith, Allan Kardec Barros, Ron Chrisley, and Vassilis Cutsuridis

Volume 658
OSTEOIMMUNOLOGY: INTERACTIONS OF THE IMMUNE AND SKELETAL SYSTEMS II
Edited by Yongwon Choi

A Continuation Order Plan is available for this series. A continuation order will bring delivery of each new volume immediately upon publication. Volumes are billed only upon actual shipment. For further information please contact the publisher.

Yongwon Choi
Editor

Osteoimmunology

Interactions of the Immune and Skeletal Systems II

Foreword by Joseph Lorenzo

Springer

Editor
Yongwon Choi
Department of Pathology & Laboratory
Medicine
Room 308, BRB II/III
421 Curie Blvd.
Philadelphia, PA 19104
USA
ychoi3@mail.med.upenn.edu

ISSN 0065-2598
ISBN 978-1-4419-1049-3 e-ISBN 978-1-4419-1050-9
DOI 10.1007/978-1-4419-1050-9
Springer New York Dordrecht Heidelberg London

Library of Congress Control Number: 2009937138

Printed on acid-free paper

Springer is part of Springer Science+Business Media (www.springer.com)

Foreword

Bone and the immune system are both complex tissues, which often interact in their function. This interaction is particularly true for the development of immune cells in the bone marrow and for the function of bone cells in health and disease. Because the two disciplines of bone biology and immunology developed independently, investigators in each do not always fully appreciate the significance of the other's system; the lack of appreciation is especially true when it comes to the function of the tissue in each discipline.

Bone has multiple functions: it provides structural integrity for the body, it is the site of hematopoiesis, and it is a storehouse for calcium and phosphorous. The immune system provides organisms with protection from invading pathogens. Multiple overlapping and interacting mechanisms have evolved to regulate both systems. Because it has become apparent that explaining the phenotype of many in vivo models with abnormal bone metabolism requires that one can no longer view the bone system in isolation. Rather, to understand its function, it must be viewed as an integrated system with the bone marrow and the immune system. Examples of recently identified interactions of bone and immune cells include the following findings: (1) that cells related to osteoblasts, which form bone, are critical regulators of the hematopoietic stem cell niche from which all blood and immune cells derive, and (2) that osteoclasts, which are the cells that resorb bone, appear to share a common origin with the myeloid precursor cells that also give rise to macrophages and myeloid dendritic cells. It has also been shown in vitro that cells, which are relatively far along in their differentiation towards antigen presenting dendritic cells, retain the ability to form mature bone resorbing osteoclasts. Finally, over the last 30 years, it has become well established that multiple soluble mediators of immune cell function, including cytokines, chemokines, and growth factors also regulate osteoblast and osteoclast activity. It is likely that immune cells and cytokines are critically responsible for the changes in bone turnover and bone mass that occur in postmenopausal osteoporosis and inflammatory conditions such as rheumatoid arthritis, periodontal disease, or inflammatory bowel disease.

The regulation of bone by hematopoietic and immune cells serves a variety of functions. It is likely that developing hematopoietic cells regulate bone turnover and maintain the marrow cavity by interacting with osteoblasts and osteoclasts during normal bone development. Conversely, during inflammatory states either locally-produced or circulating cytokines, which are the products of activated immune cells, mediate increased bone turnover and the bone pathology in diseases such as rheumatoid arthritis and inflammatory bowel disease.

The recent Second International Conference on Osteoimmunology, which occurred in Rhodes Greece on June 8th to 13, 2008, brought together a large number of investigators in this field. These proceedings represent the presentations of the invited speakers to this meeting and provide an overview of many of the most recent advances in this field. The goals of the conference were to further progress in this field and to enhance communications among scientists who study the interactions of bone and immune cells.

Farmington, Connecticut, USA Joseph Lorenzo

Contents

Contributors

Editor

Yongwon Choi Department of Pathology & Laboratory Medicine, Room 308, BRB II/III, 421 Curie Blvd., Philadelphia, PA 19104, ychoi3@mail.med.upenn.edu

Authors

Toru Akiyama Department of Orthopaedic Surgery, Faculty of Medicine, The University of Tokyo, Tokyo, Japan, akiyamat-ort@h.u-tokyo.ac.jp

Antonios O. Aliprantis Department of Medicine, Harvard Medical School, Brigham and Women's Hospital and Department of Infectious Disease and Immunology, Harvard School of Public Health, Boston, MA, USA, aaliprantis@partners.org

E.H. Allan Department of Medicine, St Vincent's Institute and University of Melbourne, Melbourne, Fitzroy, 3065, Australia, eallan@svi.edu.au

Hitoshi Amano Department of Pharmacology, School of Dentistry, Showa University, Tokyo, Japan, amano@dent.showa-u.ac.jp

Norma Andrews Pathology and Immunology, Barnes-Jewish Hospital North, Washington University School of Medicine, St. Louis, Missouri, norma.andrews@yale.edu

Atsushi Arai Division of Hard Tissue Research, Institute for Oral Science, Matsumoto Dental University, Nagano 399-0781, Japan, araiatsushi@po.mdu.ac.jp

Brendan F. Boyce The Center for Musculoskeletal Research, University of Rochester, Rochester, New York, brendan_boyce@urmc.rochester.edu

Jean Chappel Pathology and Immunology, Washington University School of Medicine, St. Louis, Missouri, USA, jchappel@wustl.edu

Wendy A. Ciovacco Department of Orthopedics and Rehabilitation, Yale University School of Medicine, New Haven, CT 06520, wendy.ciovacco@yale.edu

M.T. Gillespie Department of Medicine, Prince Henrys Institute, St Vincent's Institute and University of Melbourne, Melbourne, Clayton, 3168, Australia, matthew.gillespie@princehenrys.org

Laurie Glimcher Department of Infectious Disease and Immunology, Harvard School of Public Health and Department of Medicine, Harvard Medical School, Brigham and Women's Hospital, Boston, MA, USA, lglimche@hsph.harvard.edu

J.H. Gooi Department of Medicine, St Vincent's Institute and University of Melbourne, Melbourne, Fitzroy, 3065, Australia, jgooi@svi.edu.au

P.W.M. Ho Department of Medicine, St Vincent's Institute and University of Melbourne, Melbourne, Fitzroy, 3065, Australia, pho@svi.edu.au

Yuji Ito Pathology and Immunology, Barnes-Jewish Hospital North, Washington University School of Medicine, St. Louis, Missouri, ynth631@yahoo.co.jp

Marjo Jauhiainen Department of Pharmaceutics, University of Kuopio, Kuopio, Finland, marjo.jauhiainen@uku.fi

Dallas Jones Department of Infectious Disease and Immunology, Harvard School of Public Health, Boston, MA, djones@hsph.harvard.edu

Melissa Kacena Indiana University School of Medicine, Department of Orthopaedic Surgery, Indianapolis, IN 46202, mkacena@iupui.edu

Toshihisa Komori Unit of Basic Medical Sciences, Department of Cell Biology, Nagasaki University Graduate School of Biomedical Sciences, 1-7-1 Sakamoto, Nagasaki 852-8588, Japan, komorit@nagasaki-u.ac.jp

V. Krasnoperov Vasgene Therapeutics, Los Angeles, CA 90033, USA, valeryhome@gmail.com

Andreas Leibbrandt IMBA, Institute for Molecular Biotechnology of the Austrian Academy of Sciences, Dr. Bohr-Gasse 3, A-1030 Vienna, Austria, andreas.leibbrandt@imba.oeaw.ac.at

Joseph Lorenzo Department of Medicine, 263 Farmington Avenue, University of Connecticut Health Center, Farmington, CT 06030-1317, jlorenzo@nso2.uchc.edu

T.J. Martin Department of Medicine, St Vincent's Institute and University of Melbourne, Fitzroy, 3065, Melbourne, Australia; 2Vasgene Therapeutics, 1929 Zonal Avenue, Los Angeles, CA 90033, USA, jmartin@svi.edu.au

Koichi Matsuo Collaborative Research Resources, School of Medicine, Keio University, 160-8582 Tokyo, Japan, matsuo@sc.itc.keio.ac.jp

Toshihide Mizoguchi Division of Hard Tissue Research, Institute for Oral Science, Matsumoto Dental University, Nagano 399-0781, Japan, toshim@po.mdu.ac.jp

Hannu Mönkkönen Department of Pharmaceutics, University of Kuopio, Kuopio, Finland, hannu.monkkonen@uku.fi

Jukka Mönkkönen Department of Pharmaceutics, University of Kuopio, Kuopio, Finland, jukka.monkkonen@uku.fi

Akinori Muto Division of Hard Tissue Research, Institute for Oral Science, Matsumoto Dental University, Nagano 399-0781, Japan, mutoaki@po.mdu.ac.jp

Kozo Nakamura Department of Orthopaedic Surgery, Faculty of Medicine, The University of Tokyo, Tokyo, Japan, nakamurak-ort@h.u-tokyo.ac.jp

Keiichi I. Nakayama Department of Molecular and Cellular Biology, Medical Institute of Bioregulation, Kyushu University, Fukuoka, Japan, nakayak1@bioreg.kyushu-u.ac.jp

Josef M. Penninger IMBA, Institute of Molecular Biotechnology of the Austrian Academy of Sciences, Dr. Bohr-Gasse 3, A-1030 Vienna, Austria, josef.penninger@imba.oeaw.ac.at

Steven T. Proulx The Center for Musculoskeletal Research, University of Rochester, Rochester, New York, steven.proulx@pharma.ethz.ch

J.M.W. Quinn Department of Medicine, St Vincent's Institute and University of Melbourne, Prince Henrys Institute, Melbourne, Clayton, 3168, Australia, julian.quinn@princehenrys.org

Christopher T. Ritchlin The Center for Musculoskeletal Research, University of Rochester, Rochester, New York, christopher_ritchlin@urmc.rochester.edu

Anke J. Roelofs Bone and Musculoskeletal Research Programme, Institute of Medical Sciences, University of Aberdeen, Foresterhill, Aberdeen AB25 2ZD, UK, a.roelofs@abdn.ac.uk

Michael J. Rogers Bone and Musculoskeletal Research Programme, Institute of Medical Sciences, University of Aberdeen, Foresterhill, Aberdeen AB25 2ZD, UK, m.j.rogers@abdn.ac.uk

F. Patrick Ross Pathology and Immunology, Barnes-Jewish Hospital North, Washington University School of Medicine, St. Louis, Missouri, rossf@hss.edu

Edward M. Schwarz The Center for Musculoskeletal Research, University of Rochester Medical Center, Rochester, NY 14642, edward_schwarz@urmc.rochester.edu

N.A. Sims Department of Medicine, St Vincent's Institute and University of Melbourne, Fitzroy, 3065, Australia, nsims@svi.edu.au

Naoyuki Takahashi Division of Hard Tissue Research, Institute for Oral Science, Matsumoto Dental University, 1780 Gobara, Hiro-oka, Shiojiri-shi, Nagano 399-0781, Japan, takahashinao@po.mdu.ac.jp

Katsuhiko Takahashi Department of Biological Chemistry, School of Pharmaceutical Sciences, Showa University, Tokyo, Japan, takahask@pharm.showa-u.ac.jp

Hiroshi Takayanagi Department of Cell Signaling, Graduate School, Tokyo Medical and Dental University, Yushima 1-5-45, Bunkyo-ku, Tokyo 113-8549, Japan, taka.csi@tmd.ac.jp

Sakae Tanaka Department of Orthopaedic Surgery, Faculty of Medicine, The University of Tokyo, tanakas-ort@h.u-tokyo.ac.jp

Steve L. Teitelbaum Pathology and Immunology, Barnes-Jewish Hospital North, Washington University School of Medicine, St. Louis, Missouri, teitelbs@wustl.edu

Keith Thompson Bone and Musculoskeletal Research Programme, Institute of Medical Sciences, University of Aberdeen, Foresterhill, Aberdeen, AB25 2ZD, UK, k.thompson@abdn.ac.uk

Hidetoshi Wakeyama Department of Orthopaedic Surgery, Faculty of Medicine, The University of Tokyo, Tokyo, Japan, hwakeyama@tth-japanpost.jp

Lianping Xing The Center for Musculoskeletal Research, University of Rochester, Rochester, New York, lianping_xing@urmc.rochester.edu

Haibo Zhao Pathology and Immunology, Barnes-Jewish Hospital North, Washington University School of Medicine, St. Louis, Missouri, teitelbs@wustl.edu

The Role of Bone Marrow Edema and Lymphangiogenesis in Inflammatory-Erosive Arthritis

Edward M. Schwarz, Steven T. Proulx, Christopher T. Ritchlin, Brendan F. Boyce, and Lianping Xing

Abstract A common feature of autoimmune diseases is the perpetual production of macrophage, dendritic and/or osteoclast effector cells, which mediate parenchymal tissue destruction in end organs. In support of this, we have demonstrated previously that patients and mice with inflammatory-erosive arthritis have a marked increase in circulating CD11b+ precursor cells, which are primed for osteoclastogenesis, and that this increase in osteoclast precursors (OCPs) is due to systemically increased TNF production. From these data, we proposed a unifying hypothesis to explain these osteoimmunologic findings during the pathogenesis of inflammatory-erosive arthritis, which has three postulates: (1) myelopoiesis chronically induced by TNF has profound effects on the bone marrow and joint tissues that should be evident from a longitudinal MRI; (2) TNF alters the chemokine/chemokine receptor axis in the bone marrow to stimulate OCP release into the blood, and (3) OCP-mediated lymphangiogenesis occurs in the end organ as a compensatory mechanism to drain the inflammation and remove by-products of joint catabolism. Here, we describe our recent experimental findings that support these hypotheses and speculate on how this information can be used as diagnostic biomarkers and tools to discover novel therapies to treat patients with inflammatory-erosive arthritis.

Keywords Inflammatory arthritis · Lymphangiogenesis · In vivo imaging · 3D-MRI

1 Introduction

Based on the remarkable success of anti-TNF therapy in patients with immune-mediated inflammatory disorders (IMID) [1, 2], which include rheumatoid arthritis (RA), psoriatic arthritis (PsA), ankylosing spondylitis, Crohn's disease, and

E.M. Schwarz (✉)
The Center for Musculoskeletal Research, University of Rochester Medical Center, Rochester, New York 14642, USA
e-mail: edward_schwarz@urmc.rochester.edu

Y. Choi (ed.), *Osteoimmunology*, Advances in Experimental Medicine and Biology 658, DOI 10.1007/978-1-4419-1050-9_1, © Springer Science+Business Media, LLC 2010

psoriasis, we have pursued a unifying myelopoiesis hypothesis to explain how these drugs are so effective in seemingly unrelated diseases [3–5]. This theory posits that TNF produced in the end organ (e.g., joint, gut, and skin) acts as an endocrine factor to stimulate the proliferation and release of CD11b+ myeloid precursors from the bone marrow into the circulation where they become home to sites of inflammation and differentiate into osteoclast, macrophage, or dendritic effector cells to perpetuate autoimmune disease. In support of our theory, we have demonstrated that TNF-Tg mice [6, 7], and PsA patients [8, 9] with inflammatory-erosive arthritis have increased numbers of CD11b+ peripheral blood mononuclear cells (PBMC) that readily differentiate into bone resorbing osteoclasts; the amelioration of their disease with anti-TNF therapy is associated with a dramatic decrease in this osteoclast precursor (OPC) frequency in PBMC. To further understand the underlining etiology of IMID and develop a translational approach to test our myelopoiesis model in animals and people, we developed longitudinal MRI outcome measures to study inflammatory-erosive arthritis in mice [10–12], and we completed an exhaustive microarray analysis of mRNA isolated from CD11b+ cells in the bone marrow and blood of transgenic mice that over-express human TNF (TNF-Tg) [13] and their wild-type (WT) littermates [14–16]. These studies revealed several novel findings that provide a mechanistic understanding of how systemic TNF mediates macroscopic changes in the bone marrow to achieve perpetuating myelopoiesis and the consequences this has on inflammatory-erosive arthritis. Here, we attempt to put these osteoimmunologic findings into context with respect to how this information can be utilized to diagnosis IMID and monitor responses to therapy as well as now to pursue novel drug intervention strategies.

2 Translational MRI Outcome Measures of Inflammatory-Erosive Arthritis Identify Bone Marrow Changes and Draining Lymph Node Function as Key Biomarkers of Disease Pathogenesis

One of the greatest limitations of animal models of inflammatory-erosive arthritis is the absence of translational outcome measures that can be replicated in humans. This lack of translational outcome measures presents three problems for pre-clinical investigations of drug effects on inflammation, joint erosion, and healing [6, 17]. The problems are the following: (i) the absence of an objective assessment of disease severity prior to treatment, (ii) the extensive reliance on outcome measures that cannot be performed in humans (i.e., histologic and ex vivo molecular analyses of joint tissues during all phases of inflammatory arthritis), and (iii) inability to stratify animals randomized to drug and to placebo groups to account for the significant differences in the incidence and severity of arthritis at treatment initiation, which occurs even in genetically identical littermates. Since magnetic resonance imaging (MRI) has become the "gold standard" for the assessment of joint inflammation and damage in inflammatory arthritis in humans [18–21], and several longitudinal biomarkers of human disease have been developed for this approach (e.g., synovial inflammation and bone marrow edema) [22–27], we aimed to develop this outcome measure in mice. The approach we chose was to develop longitudinal 3-dimensional (3-D) imaging in which a custom surface coil for the mouse knee was integrated into a 3 tesla clinical MRI, utilizing pulse sequences that were compatible with human studies [12]. In these experiments, we measured inflammation using this contrast-enhanced MRI (CE-MRI) and bone erosion using in vivo microfocal computed tomography (micro-CT) to assess TNF-induced disease and its amelioration with effective therapy. The results of these studies validated these methods by demonstrating a significant relationship between changes in synovial and patellar volumes during the progression of inflammatory arthritis ($R^2 = 0.75$, $P < 0.01$). Moreover, the data demonstrated the highly variable nature of TNF-induced joint inflammation and identified the popliteal lymph node (PLN) volume as the most sensitive biomarker of ankle and knee arthritis.

3 Elucidating Bone Marrow Edema and Myelopoiesis in Murine Arthritis Using Contrast-Enhanced Magnetic Resonance Imaging

Given the dramatic effects of TNF on myelopoiesis, our model posits that macroscopic changes in the bone marrow must be evident in mice and in patients with inflammatory-erosive arthritis. In support of these gross changes, it has long been realized that arthritis is associated with enigmatic bone marrow changes that are consistent with fluid signals on MRI, and thus have been generically referred to as bone marrow edema (BME). While several retrieval studies have been performed on human tissues obtained from patients undergoing joint replacement surgery [28–31], and have grossly characterized the BME signal to be the conversion of adipocyte rich (yellow) to hematopoietic (red) marrow, the true essence of this MRI signal in terms of its (i) molecular and cellular composition, (ii) natural history, and (iii) value as a biomarker of disease pathogenesis and response to therapy has remained elusive due to the absence of an animal model and the inability to serially sample affected human joints. As expected, these BME signals were readily apparent in our mouse models [11, 12]. Thus, we developed two quantitative measures of this MRI signal to assess BME [11]. The first is normalized bone marrow intensity (NBMI), which is derived from the inherent signal intensity of the bone marrow standardized to the adjacent muscle tissue on pre-contrast MRI. The second is normalized marrow contrast enhancement (NMCE), which is enhancement of the MRI signal following intravenous administration of gadolinium contrast agent in the marrow region of interest (ROI) divided by the enhancement of signal intensity of the adjacent muscle. In our studies, WT and TNF-Tg mice were scanned from 2 to 5 months of age followed by histologic or fluorescence-activated cell sorting (FACS) analysis of marrow. We also performed efficacy studies in which TNF-Tg mice were treated with anti-TNF or placebo for 8 weeks and then studied using bimonthly MRI and histologic analysis. The results demonstrated that while NBMI values were similar in WT and TNF-Tg mice at 2 months, NBMI in WT mice steadily decreased thereafter while TNF-Tg mice retained their high values throughout life. Red to yellow marrow transformation occurred in WT but not in TNF-Tg mice as observed histologically at 5 months. Interestingly, the marrow of TNF-Tg mice treated effectively with anti-TNF therapy converted from red to yellow marrow with lower NBMI values versus placebo at 6 weeks. To confirm that the hematopoietic cells in the marrow of untreated TNF-Tg mice were OCPs, we performed FACS analysis of bone marrow, which revealed a significant correlation between NBMI values and CD11b+ monocytes ($R^2 = 0.91$, $P = 0.0028$). Perhaps of most importance to the value of this biomarker for assessing the severity of arthritis is that we were able to demonstrate thresholds values for "normal" red marrow versus pathologic BME; we also found that inflammatory marrow is highly permeable to contrast agent. Based on these findings, we conclude that BME signals in arthritic mice are caused by conversion of yellow to red marrow with associated increased myelopoiesis and increased marrow permeability. The factors that mediate these changes are currently under investigation.

Given these remarkable MRI findings and their link to TNF-induced OCP frequency (OCPF) in PBMC and focal erosions in PsA [8], we performed a clinical pilot study in which OCPF and BME were assessed in 20 patients with active erosive disease before and after etanercept therapy [9]. We found that the median OCPF decreased from 24.5 to 9% ($P = 0.04$) and 7% ($P = 0.006$) after 3 months and 6 months of treatment, respectively. Thirteen of these 20 patients completed all of the MRI evaluations. Although we found that BME decreased in 47 involved sites in these patients at 6 months, we also found that it increased in 31 other sites. Additionally, we found no correlation between OCPF, BME, and clinical parameters likely reflecting the small size of the study. These preliminary findings suggest that our myelopoiesis model of TNF-induced inflammatory-erosive arthritis is correct, but clearly these outcomes need to be assessed in an appropriately powered clinical trial. We also found that BME signals in the marrow of affected bones of PsA patients persist well after the inhibitory effects of anti-TNF therapy on OCPF are observed. Further investigation is warranted because this unexpected finding may suggest that there is ongoing subchondral inflammation or altered trabecular bone remodeling in adjacent to affected joints of patients with PsA.

4 MR Imaging and Quantification of Draining Lymph Node Function in Inflammatory Arthritis

Based on the surprising observations we made using CE-MRI of murine arthritis, which identified the PLN as a powerful biomarker of disease, we followed up on these findings with prospective studies to assess the potential of CME-MRI in our model [10]. First, we demonstrated that the volumes of PLNs of TNF-Tg mice aged 5 months are significantly larger and have greater contrast enhancement after intravenous gadolinium administration than WT control mice. This difference correlated with the abundance of dilated intranodal sinuses that were immuno-positive for the lymphatic vessel-specific marker, LYVE-1. Furthermore, we utilized dynamic CE-MRI to demonstrate differences in the kinetics of PLN enhancement between TNF-Tg and WT mice and established a LN capacity (LNcap) measurement, which is a function of both lymph node volume and CE. With this new outcome measure, we demonstrated that 5 month-old TNF-Tg mice have a 15-fold increase in lymphatic capacity over WT mice ($p < 0.001$). Interestingly, amelioration of arthritis with anti-TNF therapy resulted in a significant decrease in LNcap ($p < 0.0001$) that approached WT levels within 4 weeks of the start of treatment. However, this functional decrease was not associated with a reduction in lymphatic vessel area or numbers, which persist after therapy in both PLN and synovium. Finally, we demonstrated that there is a significant negative relationship between draining PLN function (LNcap) and synovial volume ($R^2 = 0.63$, $P = 0.01$), and that TNF-Tg mice with a lower LNcap displayed an accelerated progression of synovitis and focal joint erosion. Collectively, these results suggest that draining lymph nodes serve a function to protect joints from inflammatory arthritis, and that the initiation of joint erosion occurs when the amount of joint inflammation exceeds the maximal LNcap or that a breakdown in LNcap may occur as a result of lymph node dysfunction or "shutdown" triggered by a local immune response.

5 TNF Increases Bone Marrow Myelopoiesis by Increasing Proliferation Through Up-Regulation of c-Fms Expression on OCPs

To elucidate the mechanisms of TNF-induced increased myelopoiesis, we have focused on M-CSF and its receptor, c-Fms, because this signaling system plays a critical role in survival and proliferation of myeloid lineage cells [32]. We found that there is a two-fold increase in the percentage S-phase cells in bone marrow OCPs of TNF-Tg mice compared to their wild type littermates, suggesting increased proliferation of this cell population. In vitro, TNF induces the differentiation of bone marrow $CD11b^{+}/cFms^{-/lo}$ to $CD11b^{hi}/cFms^{hi}$ cells [33], and these $CD11b^{hi}/cFms^{hi}$ cells become highly proliferative in response to M-CSF but not to TNF [33]. Along with the recent report that TNF stimulates M-CSF production by bone marrow stromal cells [32] and our findings, we believe that two of the molecular mechanisms responsible for increased marrow myelopoiesis in IMID are TNF directly targeting OCPs to increase their c-Fms expression and stromal cells to produce M-CSF.

6 TNF Inhibits Production of SDF-1 by Bone Marrow Stromal Cells and Increases Osteoclast Precursor Mobilization from Bone Marrow to Peripheral Blood

As mentioned above, one of the central goals of our research on myelopoiesis in IMID, is the molecular mechanism by which systemic TNF mediates the release of OCPs from the bone marrow to the circulation. To address this issue, we completed a detailed investigation of the potential chemokines that could be expressed in bone marrow stroma and their cognate receptors on OCP, since these could be involved in the translocation of OCPs from the bone marrow to the blood [15]. This investigation also focused on stromal cell derived factor-1 (SDF-1/CXCL-12) and its receptor, CXCR4, because SDF-1 is the master chemokine that modulates trafficking of hematopoietic stem cells and precursors and other cell types in the bone marrow. SDF-1 is primarily produced by bone marrow stromal cells such as osteoblasts and endothelial cells [34]. Its expression has also been identified in synoviocytes in RA [35, 36], and it has been shown to be a potent chemotatic factor for OCPs [37]. We found that bone marrow stromal cells in TNF-Tg mice have reduced levels of SDF-1 expression and that TNF injection into wild type mice significantly reduced the expression of SDF-1 by stromal cells. These TNF-induced changes are associated with increase OCP frequency in the peripheral blood. Because TNF treatment did not affect the percentage of bone marrow CXCR4+ OCPs or the expression levels of CXCR4, we believe that down-regulation of SDF-1 expression by bone marrow stromal cells may be an important mechanism to explain the elevated circulating myeloid cells in IMID.

7 The Lymphatic Growth Factor, VEGF-C, is a RANKL Target Gene that Is Induced During Osteoclastogenesis and Contributes to Increased Lymphangiogenesis in Joints of Mice with Inflammatory-Erosive Arthritis

In subsequent analyses of the aforementioned microarrays [15], we looked for cytokines that were differentially expressed in OCP from TNF-Tg and WT mice [14]. To our surprise, we found that expression of vascular endothelial growth factor-C (VEGF-C), whose primary function is considered to be regulation of lymphangiogenesis [38], was markedly increased in TNF-Tg mice OCPs. While it was known that VEGF-C is highly expressed by numerous cancer cells, inflammatory macrophages, and synoviocytes [39–41] the factors that control VEGF-C expression have not been studied in detail. Therefore, we investigated the effects of TNF, IL-1, and RANKL on the expression VEGF family members in OCPs. We found that RANKL increased *vegf-c* mRNA expression 12-fold while TNF and IL-1 increased its expression only 5-fold. Moreover, this effect of RANKL was specific for VEGF-C, as this was the only VEGF family member that was increased. These data provided the first evidence that osteoclastogenesis and lymphangiogenesis are concomitant events and suggest that these physiological events may be linked in the pathogenesis of inflammatory-erosive arthritis.

In order to examine the physiological significance of this VEGF-C induction in inflammatory arthritis, we assessed the effects of TNF on VEGF-C expression and lymphangiogenesis in TNF-Tg mice and in mice with serum-induced arthritis [16].

We used immunostaining and MRI to quantify lymphatic vessels and the volumes of synovium and draining lymph nodes. We found that TNF stimulated VEGF-C expression by OCPs in a NF-κB-dependent manner. Moreover, OCPs from joints of TNF-Tg mice express high levels of VEGF-C, and this VEGF-C had autocrine effects to induce osteoclastic bone resorption through activation of the c-Src signaling pathway. Lymphatic vessel numbers and size were markedly increased in joint sections of TNF-Tg mice and in mice with serum-induced arthritis. Additionally, the severity of synovitis correlated with draining lymph node size. This led us to conclude that TNF induces OCPs to produce VEGF-C, which significantly increased lymphangiogenesis in joints of arthritic mice, and supports our emerging hypothesis that the lymphatic system may play an important role in the pathogenesis of inflammatory arthritis.

8 Conclusions

Based on our recent findings and the established literature, we now propose a comprehensive model to explain the mechanisms whereby TNF produced in arthritic joints stimulates OCP production and release from the bone marrow to propagate a vicious cycle, which is manifested as inflammatory-erosive arthritis (Fig. 1). The pathogenesis commences with an unknown event that leads to chronic TNF over expression in affected joints. TNF has endocrine effects on the bone marrow where it stimulates M-CSF production by stromal cells and differentiation of

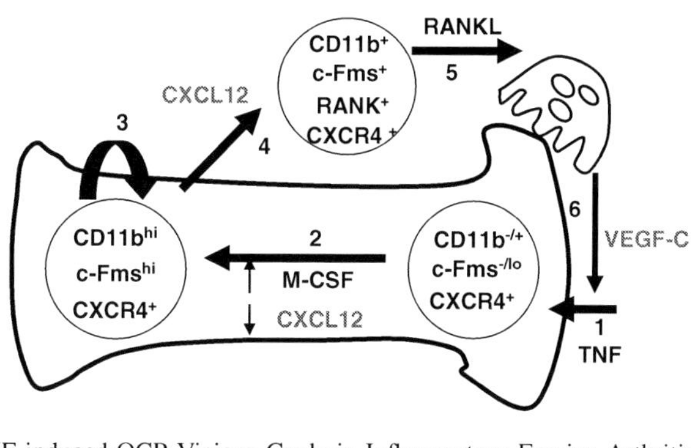

Fig. 1 The TNF-induced OCP Vicious Cycle in Inflammatory-Erosive Arthritis. (*1*) The pathogenesis commences with an unknown event that leads to chronic TNF over expression in the joint. (*2*) This TNF has endocrine effects on the bone marrow where it stimulates M-CSF production by stromal cells and differentiation of hematopoietic progenitors to CD11b^{hi}/cFms^{hi} OCPs. (*3*) M-CSF stimulates the proliferation of CD11b^{hi}/cFms^{hi} OCP, which (*4*) leads to their migration from the bone marrow to the blood via a mechanism that is involves the down-regulation of stromal cell derived factor -1 (CXCL12). (*5*) PBMC/OCPs are recruited into the inflamed synovium, and are stimulated by RANKL to become bone resorbing osteoclasts. (*6*) This RANKL stimulation also induces the expression of lymphangiogenic factor, VEGF-C and perhaps other yet to be defined angiogenic and lymphangiogenic factors that complete the vicious cycle through paracrine/endocrine effects on inflammation and myelopoiesis

hematopoietic progenitors from $CD11b^{hi}/cFms^{-/lo}$ to $CD11b^{hi}/cFms^{hi}$ OCPs. These bone marrow changes lead to increased proliferation of $CD11b^{hi}/cFms^{hi}$ OCPs in response to M-CSF and enhanced bone marrow myelopoiesis. In the meantime, TNF reduces SDF-1 production by bone marrow stromal cells and increases the mobilization of OCPs from bone marrow to the peripheral blood. Joint inflammation then recruits circulating OCPs to the inflamed synovium where these cells are stimulated by RANKL and TNF to become bone resorbing osteoclasts and also produce angiogenic and lymphangiogenic factors such as VEGF-C to stimulate vasculogenesis.

Based on TNF's pleiotropic inflammatory effects on virtually all cell types in the body, it is not surprising that it is a central factor in autoimmune diseases. However, its dominant role in perpetuating the release of myeloid effector cells from the bone marrow now appears to be a common link in IMID. Thus, a current challenge is to further understand the molecular mechanisms of this osteoimmunologic phenomenon, which includes elucidation of the factors that mediate TNF-induced yellow to red bone marrow conversion, and the generation of novel interventions that are less immunosuppressive and more cost-effective than anti-TNF therapy.

References

1. Appel, H., Kuhne, M., Spiekermann, S., Kohler, D., Zacher, J., Stein, H. et al. (2006). Immunohistochemical analysis of hip arthritis in ankylosing spondylitis: evaluation of the bone-cartilage interface and subchondral bone marrow. *Arthritis Rheum,* **54**(6):1805–1813.
2. Appel, H., Loddenkemper, C., Grozdanovic, Z., Ebhardt, H., Dreimann, M., Hempfing, A. et al. (2006). Correlation of histopathological findings and magnetic resonance imaging in the spine of patients with ankylosing spondylitis. *Arthritis Res Ther,* **8**(5):R143
3. Anandarajah, A.P., Schwarz, E.M., Totterman, S., Monu, J., Feng, C.Y., Shao, T. et al. (2008). The effect of etanercept on osteoclast precursor frequency and enhancing bone marrow oedema in patients with psoriatic arthritis. *Ann Rheum Dis,* **67**(3):296–301.
4. Benton, N., Stewart, N., Crabbe, J., Robinson, E., Yeoman, S., & McQueen, F.M. (2004). MRI of the wrist in early rheumatoid arthritis can be used to predict functional outcome at 6 years. *Ann Rheum Dis,* **63**(5):555–561.
5. Boyce, B.F., Schwarz, E.M., & Xing, L. (2006). Osteoclast precursors: cytokine-stimulated immunomodulators of inflammatory bone disease. *Curr Opin Rheumatol,* **18**(4):427–432.
6. Bradfield, P.F., Amft, N., Vernon-Wilson, E., Exley, A.E., Parsonage, G., Rainger, G.E. et al. (2003). Rheumatoid fibroblast-like synoviocytes overexpress the chemokine stromal cell-derived factor 1 (CXCL12), which supports distinct patterns and rates of CD4+ and CD8+ T cell migration within synovial tissue. *Arthritis Rheum,* **48**(9):2472–2482.
7. Conaghan, P.G., McQueen, F.M., Peterfy, C.G., Lassere, M.N., Ejbjerg, B., Bird, P. et al. (2005). The evidence for magnetic resonance imaging as an outcome measure in proof-of-concept rheumatoid arthritis studies. *J Rheumatol,* **32**(12):2465–2469.
8. Dohn, U.M., Ejbjerg, B.J., Court-Payen, M., Hasselquist, M., Narvestad, E., Szkudlarek, M. et al. (2006). Are bone erosions detected by magnetic resonance imaging and ultrasonography true erosions? A comparison with computed tomography in rheumatoid arthritis metacarpophalangeal joints. *Arthritis Res Ther,* **8**(4):R110.
9. Feldmann, M., & Maini, R.N. (2003). Lasker Clinical Medical Research Award. TNF defined as a therapeutic target for rheumatoid arthritis and other autoimmune diseases. *Nat Med,* **9**(10):1245–1250.

10. Haavardsholm, E.A., Ostergaard, M., Ejbjerg, B.J., Kvan, N.P., Uhlig, T.A., Lilleas, F.G. et al. (2005). Reliability and sensitivity to change of the OMERACT rheumatoid arthritis magnetic resonance imaging score in a multireader, longitudinal setting. *Arthritis Rheum,* **52**(12):3860–3867.
11. Jimenez-Boj, E., Nobauer-Huhmann, I., Hanslik-Schnabel, B., Dorotka, R., Wanivenhaus, A.H., Kainberger, F. et al. (2007). Bone erosions and bone marrow edema as defined by magnetic resonance imaging reflect true bone marrow inflammation in rheumatoid arthritis. *Arthritis Rheum,* **56**(4):1118–1124.
12. Jimenez-Boj, E., Redlich, K., Turk, B., Hanslik-Schnabel, B., Wanivenhaus, A., Chott, A. et al. (2005). Interaction between synovial inflammatory tissue and bone marrow in rheumatoid arthritis. *J Immunol,* **175**(4):2579–2588.
13. Keffer, J., Probert, L., Cazlaris, H., Georgopoulos, S., Kaslaris, E., Kioussis, D. et al. (1991). Transgenic mice expressing human tumour necrosis factor: a predictive genetic model of arthritis. *Embo J,* **10**(13):4025–4031.
14. Kirkham, B.W., Lassere, M.N., Edmonds, J.P., Juhasz, K.M., Bird, P.A., Lee, C.S. et al. (2006). Synovial membrane cytokine expression is predictive of joint damage progression in rheumatoid arthritis: a two-year prospective study (the DAMAGE study cohort). *Arthritis Rheum,* **54**(4):1122–1131.
15. Kitaura, H., Zhou, P., Kim, H.J., Novack, D.V., Ross, F.P., & Teitelbaum, S.L.(2005). M-CSF mediates TNF-induced inflammatory osteolysis. *J Clin Invest,* **115**(12):3418–3427
16. Kuek, A., Hazleman, B.L., & Ostor, A.J. (2007). Immune-mediated inflammatory diseases (IMIDs) and biologic therapy: a medical revolution. *Postgrad MedJ,* **83**(978):251–260.
17. Li, P., Schwarz, E.M., O'Keefe, R.J., Ma, L., Boyce, B.F., & Xing, L. (2004). RANK signaling is not required for TNFalpha-mediated increase in CD11(hi) osteoclast precursors but is essential for mature osteoclast formation in TNFalpha-mediated inflammatory arthritis. *J Bone Miner Res,* **19**(2):207–213.
18. Li, P., Schwarz, E.M., O'Keefe, R.J., Ma, L., Looney, R.J., Ritchlin, C.T. et al. (2004). Systemic tumor necrosis factor alpha mediates an increase in peripheral CD11bhigh osteoclast precursors in tumor necrosis factor alpha-transgenic mice. *Arthritis Rheum,* **50**(1):265–276.
19. Martinez-Martinez, M.U., Cuevas-Orta, E., Reyes-Vaca, G., Baranda, L., Gonzalez-Amaro, R., & Abud-Mendoza, C. (2007). Magnetic resonance imaging in patients with rheumatoid arthritis with complete remission treated with disease-modifying antirheumatic drugs or anti-tumour necrosis factor alpha agents. *Ann Rheum Dis,* **66**(1):134–135.
20. Maruyama, K., Asai, J., Ii, M., Thorne, T., Losordo, D.W., & D'Amore, P.A. (2007). Decreased macrophage number and activation lead to reduced lymphatic vessel formation and contribute to impaired diabetic wound healing. *Am J Pathol,* **170**(4):1178–1191
21. McGonagle, D., Conaghan, P.G., O'Connor, P., Gibbon, W., Green, M., Wakefield, R. et al. (1999). The relationship between synovitis and bone changes in early untreated rheumatoid arthritis: a controlled magnetic resonance imaging study. *Arthritis Rheum,* **42**(8):1706–1711.
22. McQueen, F.M. (2000). Magnetic resonance imaging in early inflammatory arthritis: what is its role? *Rheumatology,* **39**(7):700–706.
23. McQueen, F.M., & Ostendorf, B. (2006). What is MRI bone oedema in rheumatoid arthritis and why does it matter? *Arthritis Res Ther,* **8**(6):222.
24. Oliver, G., & Detmar, M. (2002). The rediscovery of the lymphatic system: old and new insights into the development and biological function of the lymphatic vasculature. *Genes Dev,* **16**(7):773–783.
25. Ostergaard, M., Hansen, M., Stoltenberg, M., Gideon, P., Klarlund, M., Jensen, K.E., & Lorenzen, I. (1999). Magnetic resonance imaging-determined synovial membrane volume as a marker of disease activity and a predictor of progressive joint destruction in the wrists of patients with rheumatoid arthritis. *Arthritis Rheum,* **42**(5):918–929.
26. Pablos, J.L., Santiago, B., Galindo, M., Torres, C., Brehmer, M.T., Blanco, F.J. et al. (2003). Synoviocyte-derived CXCL12 is displayed on endothelium and induces angiogenesis in rheumatoid arthritis. *J Immunol,* **170**(4):2147–2152.

27. Ponomaryov, T., Peled, A., Petit, I., Taichman, R.S., Habler, L., Sandbank, J. et al. (2000). Induction of the chemokine stromal-derived factor-1 following DNA damage improves human stem cell function. *J Clin Invest,* **106**(11):1331–1339.
28. Proulx, S.T., Kwok, E., You, Z., Beck, C.A., Shealy, D.J., Ritchlin, C.T. et al. (2007). MRI and quantification of draining lymph node function in inflammatory arthritis. *Ann N Y Acad Sci,* **1117**:106–123.
29. Proulx, S.T., Kwok, E., You, Z., Papuga, M.O., Beck, C.A., Shealy, D.J. et al. (2008). Elucidating bone marrow edema and myelopoiesis in murine arthritis using contrast-enhanced magnetic resonance imaging. *Arthritis Rheum,* **58**(7):2019–2029.
30. Proulx, S.T., Kwok, E., You, Z., Papuga, M.O., Beck, C.A., Shealy, D.J. et al. (2007). Longitudinal assessment of synovial, lymph node, and bone volumes in inflammatory arthritis in mice by in vivo magnetic resonance imaging and microfocal computed tomography. *Arthritis Rheum,* **56**(12):4024–4037.
31. Ritchlin, C.T., Haas-Smith, S.A., Li, P., Hicks, D.G., & Schwarz, E.M. (2003). Mechanisms of TNF-alpha- and RANKL-mediated osteoclastogenesis and bone resorption in psoriatic arthritis. *J Clin Invest,* **111**(6):821–831.
32. Salven, P., Lymboussaki, A., Heikkila, P., Jaaskela-Saari, H., Enholm, B., Aase, K. et al. (1998). Vascular endothelial growth factors VEGF-B and VEGF-C are expressed in human tumors. *Am J Pathol,* **153**(1):103–108.
33. Schwarz, E.M., Looney, R.J., Drissi, M.H., O'Keefe, R.J., Boyce, B.F., Xing, L., & Ritchlin, C.T. (2006). Autoimmunity and bone. *Ann N Y Acad Sci,* **1068**:275–283.
34. Shealy, D.J., Wooley, P.H., Emmell, E., Volk, A., Rosenberg, A., Treacy, G. et al. (2002). Anti-TNF-alpha antibody allows healing of joint damage in polyarthritic transgenic mice. *Arthritis Res,* **4**(5):R7.
35. Su, J.L., Yen, C.J., Chen, P.S., Chuang, S.E., Hong, C.C., Kuo, I.H. et al. (2007). The role of the VEGF-C/VEGFR-3 axis in cancer progression. *Br J Cancer,* **96**(4):541–545.
36. Yu, X., Huang, Y., Collin-Osdoby, P., & Osdoby, P. (2003). Stromal cell-derived factor-1 (SDF-1) recruits osteoclast precursors by inducing chemotaxis, matrix metalloproteinase-9 (MMP-9) activity, and collagen transmigration. *J Bone Miner Res,* **18**(8):1404–1418.
37. Xing, L., Schwarz, E.M., & Boyce, B.F. (2005). Osteoclast precursors, RANKL/RANK, and immunology. *Immunol Rev,* **208**:19–29.
38. Yao, Z., Li, P., Zhang, Q., Schwarz, E.M., Keng, P., Arbini, A. et al. (2006). Tumor Necrosis Factor-{alpha} Increases Circulating Osteoclast Precursor Numbers by Promoting Their Proliferation and Differentiation in the Bone Marrow through Up-regulation of c-Fms Expression. *J Biol Chem,* **281**(17):11846–11855.
39. Zhang, Q., Guo, R., Lu, Y., Zhao, L., Zhou, Q., Schwarz, E.M. et al. (2008). VEGF-C, a Lymphatic Growth Factor, Is a RANKL Target Gene in Osteoclasts That Enhances Osteoclastic Bone Resorption through an Autocrine Mechanism. *J Biol Chem,* **283**(19):13491–13499.
40. Zhang, Q., Guo, R., Schwarz, E.M., Boyce, B.F., & Xing, L. (2008). TNF inhibits production of SDF-1 by bone stromal cells and increases osteoclast precursor mobilization from bone marrow to peripheral blood. *Arthritis Res Ther,* **10**(2):R37.
41. Zhang, Q., Lu, Y., Proulx, S., Guo, R., Yao, Z., Schwarz, E.M. et al. (2007). Increased lymphangiogenesis in joints of mice with inflammatory arthritis. *Arthritis Res Ther*, **9**(6):R118.

Activation of γδ T Cells by Bisphosphonates

Keith Thompson, Anke J. Roelofs, Marjo Jauhiainen, Hannu Mönkkönen, Jukka Mönkkönen, and Michael J. Rogers

Abstract After decades of successful clinical use, the exact molecular mechanisms by which the anti-resorptive bisphosphonate drugs (BPs) exert their effects are now being revealed. In addition to their anti-resorptive effects, it is now apparent that nitrogen-containing BPs (N-BPs) have immunomodulatory properties. Specifically, these drugs activate immune cells called gamma, delta T lymphocytes. In this chapter we discuss the mechanism of gamma, delta T cell activation by N-BPs and propose that N-BPs may provide a safe and effective means for manipulating gamma,delta T cell activity in future immunotherapeutic approaches.

Keywords Bisphosphonate · Gamma, delta T cell · Monocytes · Osteoporosis · Mevalonate pathway

1 Bisphosphonates

Bisphosphonates (BPs) are a class of drugs successfully used to treat a wide variety of diseases characterized by excessive osteoclast-mediated bone resorption, such as tumour-associated osteolysis, Paget's disease of bone and post-menopausal osteoporosis. BPs consist of a common geminal carbon atom linked to two phosphonate groups, to yield a P–C–P structure and, as such, BPs are considered to be analogues of naturally occurring pyrophosphate (P–O–P). However, unlike pyrophosphate, the P–C–P backbone of BPs confers a remarkable resilience to enzymatic hydrolysis. The incorporation of a carbon, rather than oxygen, atom in the P–C–P structure also allows the addition of two further side chains, thus allowing a great number of possible variations in molecular structure (Fig. 1), which influence both the potency and pharmacology of these drugs.

K. Thompson (✉)
Bone and Musculoskeletal Research Programme, Institute of Medical Sciences, University of Aberdeen, Foresterhill, Aberdeen AB25 2ZD, UK
e-mail: k.thompson@abdn.ac.uk

Y. Choi (ed.), *Osteoimmunology*, Advances in Experimental Medicine and Biology 658, DOI 10.1007/978-1-4419-1050-9_2,

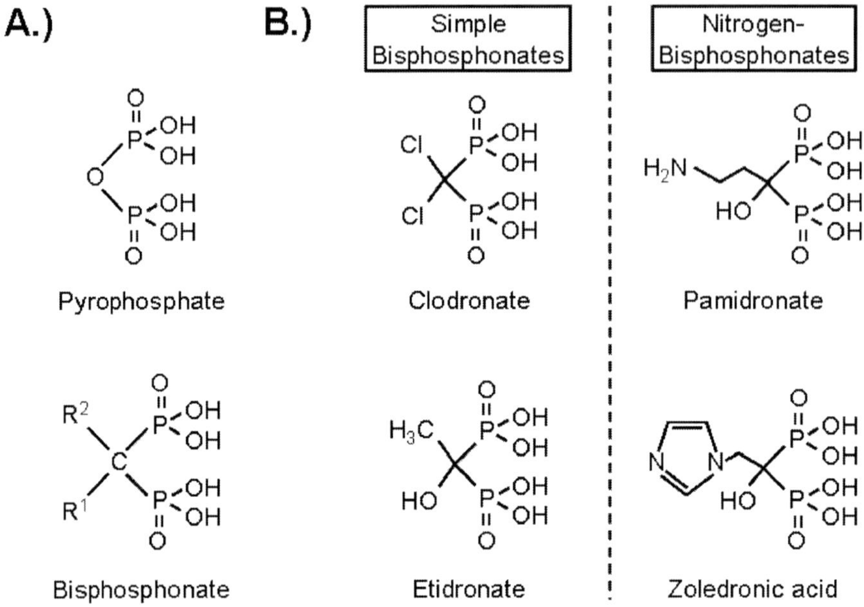

Fig. 1 (**A**) The structure of pyrophosphate and a geminal bisphosphonate. R^1 and R^2 indicate the positions of the variable side chains. (**B**) Structures of bisphosphonates (shown as the protonated bisphosphonic acid forms)

The two phosphonate groups present in BPs also endow these compounds with a high affinity for divalent ions, such as calcium. Due to the abundance of calcium in bone mineral, one of the key pharmacological features of these drugs is rapid sequestration by the skeleton. Binding of BPs to bone mineral is crucial for the relatively selective targeting of osteoclasts by these drugs, since these cells are the only cell type currently known that can liberate bound BPs from the bone surface during the resorptive process [6]. During the process of bone resorption, osteoclasts secrete protons to acidify the resorption pit, which decreases the affinity of BPs for calcium ions and causes release of the BP from the bone surface. BPs are then internalized by the osteoclast during a process known as fluid-phase endocytosis [30]. Once BPs gain entry into the osteoclast, they inhibit osteoclast activity by one of two mechanisms, based on the molecular structure of the BP.

2 Mechanism of Action of Bisphosphonates

Simple BPs lacking a nitrogen moiety, such as clodronate (CLO) and etidronate (ETI), are metabolized intracellularly into non-hydrolyzable cytotoxic analogs of ATP, which accumulate in the osteoclast and trigger apoptosis [11]. In contrast, the more potent nitrogen-BPs (N-BPs), such as pamidronate (PAM) and zoledronic acid (ZOL), inhibit osteoclast function by acting as potent inhibitors of the enzyme

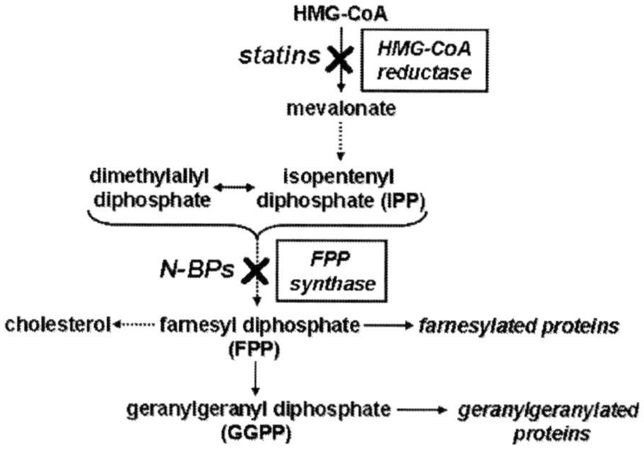

Fig. 2 Schematic representation of the cholesterol biosynthetic pathway. The sites of inhibition by statins and nitrogen-bisphosphonates (N-BPs) are indicated

farnesyl diphosphate (FPP) synthase [2, 9, 33] in the cholesterol (or mevalonate) biosynthetic pathway (Fig. 2). FPP synthase activity is essential for the maintenance of an intracellular pool of isoprenoid lipids such as FPP and geranylgeranyl diphosphate (GGPP), which are required for the post-translational modification (isoprenylation) of small GTP-binding proteins such as Rho, Rac and Cdc42. Isoprenylation of these signalling proteins is crucial for the correct sub-cellular localization and function of these GTPases [35]. As a result, inhibition of FPP synthase by N-BPs results in defective regulation of processes reliant on GTPase activity such as cytoskeletal rearrangement [23, 26] and vesicular trafficking [36] in osteoclasts (for review see [5]).

Oral administration of BPs is generally well tolerated, and their potent inhibitory effect on excessive osteoclast activity has resulted in the successful widespread clinical use of these compounds for over 30 years. The profound anti-resorptive effect of a once-yearly infusion of zoledronic acid [3], or more frequent infusions of ibandronate [8], has demonstrated that the intravenous administration of BPs may be an attractive therapeutic alternative to oral administration. The most common adverse effect of intravenous N-BP administration, which is not generally observed with oral dosing, is the development of a self-limiting, flu-like syndrome called the acute-phase reaction (APR) [1, 25], which typically occurs in one-third of patients. Despite this syndrome being first reported over 20 years ago, the molecular events triggering this adverse event of N-BPs are only now becoming apparent.

3 γδ T Cells and the Acute Phase Response to Nitrogen-Bisphosphonates

The molecular mechanism underlying the acute-phase response (APR) was first revealed by Kunzmann et al. in 1999. Their seminal study revealed that in multiple myeloma patients treated with PAM, the patients who developed an APR had increased circulating levels of γδ T cells up to 28 days following the infusion [16]. Furthermore, the severity of the APR, as assessed by increased body temperature, correlated to the extent of the increase in γδ T cell numbers observed in these patients. Subsequent studies revealed that only the predominant subset of γδ T cells in peripheral blood, Vγ9Vδ2 T cells, were activated by N-BPs [15]. Thus, it was apparent that the N-BP-induced Vγ9Vδ2 T cell activation, and subsequent release of pro-inflammatory cytokines such as TNFα, IL-6 and IFNγ [25, 24, 29], played a crucial role in the APR; however, the exact mechanism by which N-BPs activated Vγ9Vδ2 T cells remained to be elucidated.

γδ T cells are non-conventional T cells that recognize non-peptide antigens without the need for conventional MHC class presentation, but they also exhibit characteristics of natural killer (NK) cells and cytotoxic T cells, since they can recognize and kill target cells through Fas-ligand- and perforin/granzyme B-dependent mechanisms. In humans, γδ T cells comprise only a minor proportion (1–10%) of $CD3^+$ T cells in peripheral blood, and the majority (50–90%) belong to the Vγ9Vδ2 subset (also termed Vγ2Vδ2) [13]. While the exact roles of γδ T cells in humans have yet to be determined, these cells are thought to play an important role in tumour surveillance [17]. Furthermore, they have potent anti-microbial actions, and due to their presence in various epithelial tissues, they may present an important first line of defence to invading pathogens [13].

Until recently, N-BPs were thought to activate Vγ9Vδ2 T cells by acting as agonists for the Vγ9Vδ2-TCR [7]. However, using a Burkitt's lymphoma cell line (Daudi), Gober et al. demonstrated that ZOL required internalization by Daudi cells to induce activation of γδ T cell clones, thereby arguing against a direct agonistic effect of ZOL on the Vγ9Vδ2-TCR [12]. It is now apparent that N-BPs *indirectly* activate Vγ9Vδ2 T cells through the intracellular inhibition of the molecular target of the N-BPs, FPP synthase. FPP synthase inhibition, as well as causing a depletion of the downstream metabolites FPP and GGPP (and the consequent inhibition of protein prenylation), also induces the accumulation of the substrates of this enzyme, IPP and dimethylallyl diphosphate (DMAPP) [32] (see Fig. 2), which are both agonists for the Vγ9Vδ2-TCR [28]. Gober et al. demonstrated that the stimulatory effect of ZOL on Vγ9Vδ2 T cell activation by tumor cells was due to the accumulation of these upstream mevalonate pathway intermediates in the tumor cells [12]. At the same time, our own parallel studies using peripheral blood mononuclear cell (PBMC) cultures, revealed that the stimulatory capacity of N-BPs for inducing Vγ9Vδ2 T cell activation was determined by the potency of the N-BP for inhibiting FPP synthase [31] (Fig. 3). Furthermore, N-BP-induced Vγ9Vδ2 T cell activation could be blocked by a statin, which inhibits the upstream enzyme HMG-CoA reductase, through a mechanism most likely involving inhibition of IPP/DMAPP synthesis [12, 14, 31]. We also demonstrated that this inhibitory effect of statins was selective for N-BP-induced Vγ9Vδ2 T cell activation, since statins did not inhibit Vγ9Vδ2 T cell activation induced by well characterized Vγ9Vδ2-TCR agonists such as IPP (unpublished observations), or the synthetic Vγ9Vδ2-TCR agonist, bromohydrin pyrophosphate [10, 31].

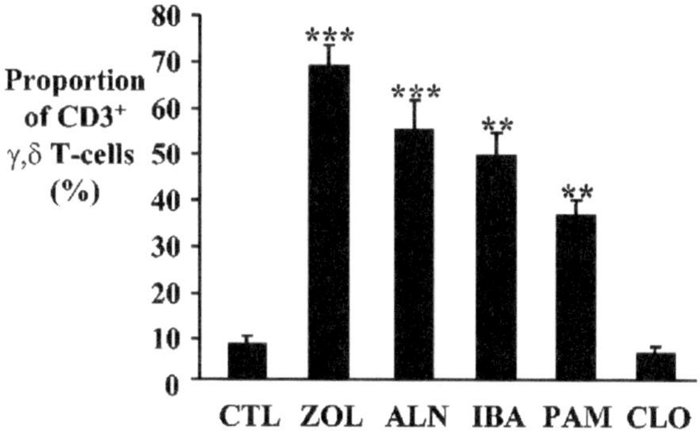

Fig. 3 The effect of N-BPs on γδ T cell proliferation. PBMCs were cultured for 7 days with 1 μM ZOL, ALN, IBA, PAM, or CLO in the presence of rhIL-2. PBMCs were dual-stained with anti-CD3-FITC and anti-pan-γδ-TCR-PC5 antibodies before FACS analysis of the T cell-gated population. Data shown are the mean of experiments with PBMCs from four independent donors ± S.E.M. $^{***}p < 0.001$; $^{**}p < 0.01$. Reproduced from Thompson and Rogers [31], with permission of the American Society for Bone and Mineral Research

4 Triggering of the Acute-Phase Response to Nitrogen-Bisphosphonates In Vivo

In N-BP-treated PBMC cultures IPP/DMAPP accumulation undoubtedly plays a crucial role in triggering Vγ9Vδ2 T cell activation, but it was unknown which cell types present in peripheral blood were directly targeted by N-BPs (and hence, accumulate IPP/DMAPP). Following a typical intravenous infusion of ZOL, the peak plasma concentration reaches ~1 μM and is only maintained in the peripheral circulation for ~2 h [4], due to a combination of both rapid renal excretion and sequestration by the skeleton. Thus, in order to trigger an APR, ZOL must be internalized rapidly by peripheral blood cells.

We have previously shown that highly endocytic cells such as macrophages are relatively sensitive to N-BPs in vitro [6]. By using a fluorescently labelled N-BP analog, we revealed that N-BPs are internalized into macrophages and osteoclasts predominantly by the process of fluid-phase endocytosis [30] (Fig. 4). Therefore, it is likely that the triggering of an APR results from the rapid uptake of N-BP by a cell population that has high endocytic activity present in peripheral blood. Monocytes (present in PBMC cultures) have previously been shown to be crucial for N-BP-induced Vγ9Vδ2 T cell activation [20], although the role of monocytes in this process was at that time thought to involve presentation of N-BP as antigen to Vγ9Vδ2 T cells and/or provision of a crucial co-stimulatory role, thereby decreasing the threshold of Vγ9Vδ2 T cell activation.

Our recent studies have clarified the role of monocytes in Vγ9Vδ2 T cell activation in vitro. Using our fluorescent N-BP analog-based approach [6, 30], we have shown that N-BPs are selectively internalized by peripheral blood monocytes in human PBMC cultures, with no detectable uptake into B- or T-lymphocytes, at

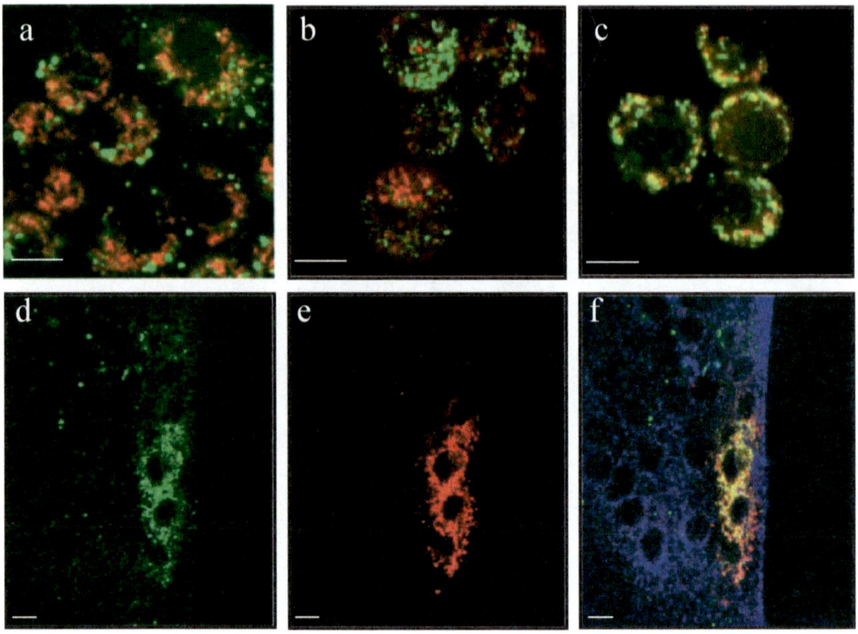

Fig. 4 Fluorescently labelled bisphosphonate (alendronate-Alexa Fluor 488: AF-ALN) is internalized by fluid-phase endocytosis in J774.2 monocyte-macrophage cells and in rabbit osteoclasts. J774.2 cells (**a–c**) were incubated with (**a**) 100 μM AF-ALN + 1 μg/ml wheat germ agglutinin-Alexa Fluor 633 (a marker of adsorptive endocytosis); (**b**) 100 μM AF-ALN + 20 μg/ml transferrin-Alexa Fluor 633 (a marker of receptor-mediated endocytosis); or (**c**) 100 μM AF-ALN + 250 μg/ml TAMRA-dextran (a marker of fluid-phase endocytosis). Rabbit osteoclasts (**d–f**) were treated with (**d**) 100 μM AF-ALN (green); (**e**) 250 μg/ml TAMRA-dextran (red); or (**f**) 100 μM AF-ALN + TAMRA-dextran + transferrin-Alexa Fluor 633 (blue), for 6 h then fixed and analyzed by confocal microscopy (scale bar, 10 μm). In both J774.2 cells and rabbit osteoclasts co-localization of AF-ALN is only observed with fluorescently-labelled dextran, indicating fluid-phase endocytosis is the major route by which AF-ALN gains entry into these cell types. Reproduced from Thompson et al. [30], with permission of the American Society for Pharmacology and Experimental Therapeutics

low and clinically relevant concentrations [22]. In addition, only $CD14^+$ monocytes in these cultures efficiently internalized FITC-dextran, a marker of fluid-phase endocytosis, and N-BP uptake co-localized with uptake of FITC-dextran, suggesting that fluid-phase endocytosis is the major mechanism by which monocytes internalize N-BPs. Furthermore, using magnetic bead-separation approaches, combined with HPLC-ESI-MS, we have also revealed that treatment of human PBMC cultures with a pharmacologically relevant concentration (1 μM) and duration (2 h) of ZOL induced a detectable accumulation of IPP/DMAPP only in the $CD14^+$ fraction. Importantly, we also demonstrated that statins prevent ZOL-induced IPP/DMAPP accumulation [21]. This raises the intriguing possibility that statins may potentially diminish or even prevent the APR to N-BPs in vivo.

5 Further Questions

Currently, one aspect of the APR to N-BPs that is poorly understood is the development of symptoms following the initial infusion, but the subsequent lack of symptoms with repeated N-BP infusions. Similarly, decreased responsiveness of Vγ9Vδ2 T cells to repeated treatment with the synthetic Vγ9Vδ2-TCR agonist BrHPP has been reported in cynomolgus monkeys [27]. This suggests that there is some inherent regulatory mechanism(s) that limit(s) the response following repeated challenges with Vγ9Vδ2 T cell activators. Furthermore, it is at present still unclear why only approximately one-third of patients develop flu-like symptoms of an APR following the first infusion, since the majority ($\geq$90%) of blood samples isolated from healthy donors demonstrate reactive Vγ9Vδ2 T cells when stimulated with clinically relevant concentrations of N-BPs in vitro (our unpublished observations and [19, 34]). The mechanism underlying this diminished responsiveness of Vγ9Vδ2 T cells to N-BPs in vivo remains to be clarified, but may involve regulatory T cells (T_{regs}), $CD4^+/CD25^{hi}/FoxP3^+$ T cells, which have recently been demonstrated to suppress IFNγ production by antigen-stimulated Vγ9Vδ2 T cells in vitro [18]. Should T_{regs} be found to inhibit Vγ9Vδ2 T cell activation in vivo, strategies to minimize the inhibitory effects of T_{regs} on Vγ9Vδ2 T cells may allow for more effective deployment of γδ T cell-based therapies for the treatment of lymphoid malignancies and other types of cancer, as well as bacterial and viral infections (Fig. 5).

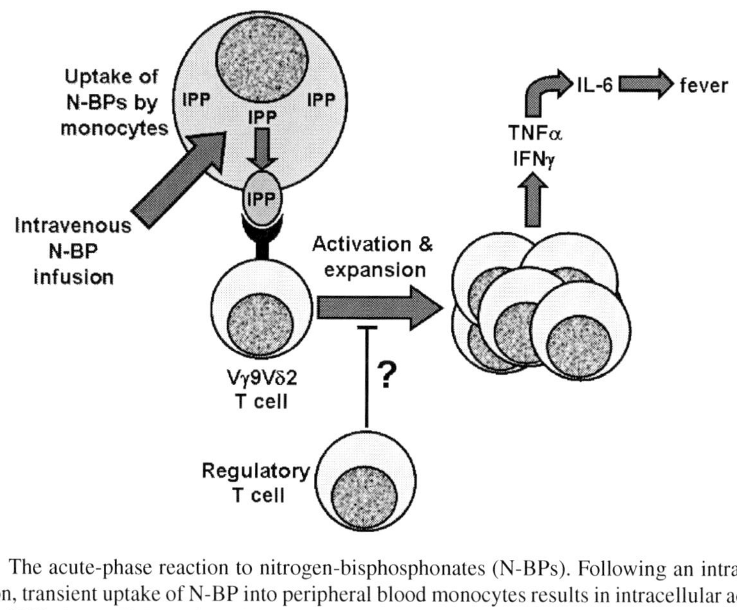

Fig. 5 The acute-phase reaction to nitrogen-bisphosphonates (N-BPs). Following an intravenous infusion, transient uptake of N-BP into peripheral blood monocytes results in intracellular accumulation of IPP due to FPP synthase inhibition. Recognition of IPP by Vγ9Vδ2 T cells triggers their activation and expansion, resulting in the release of pro-inflammatory cytokines that cause the flu-like symptoms of the acute-phase reaction. The relevance of regulatory T cell-mediated inhibitory effects on N-BP-induced Vγ9Vδ2 T cell activation in vivo is currently unknown.

6 Conclusion

The molecular mechanisms responsible for activation of Vγ9Vδ2 T cells by N-BPs, and their role in the APR, have only recently become clear, despite the identification of this syndrome more than 20 years ago. Crucially, this adverse event of intravenous N-BP therapy has unwittingly revealed the potent stimulatory effects of N-BPs on Vγ9Vδ2 T cells. This may allow the use of these compounds in future γδ T cell-based immunomodulatory strategies to elicit anti-tumor effects, particularly in conditions such as lymphoid malignancies, which are often associated with osteolytic disease. Due to the relative safety, potency and long-term stability of N-BPs such as ZOL, these agents represent ideal candidates for activating Vγ9Vδ2 T cells in the clinical setting, particularly where anti-resorptive effects are beneficial. While many questions remain about γδ T cell biology, such as their possible regulation by T_{regs}, the therapeutic manipulation of γδ T cells has great potential and may form the basis of novel anti-tumor, anti-bacterial and anti-viral treatment strategies in the future.

References

1. Adami, S., Bhalla, A.K., Dorizzi, R. et al. (1987). The acute-phase response after bisphosphonate administration. *Calcif Tiss Int*, 41, 326–331.
2. Bergstrom, J.D., Bostedor, R.G., Masarachia, P.J., Reszka, A.A., & Rodan, G. (2000). Alendronate is a specific, nanomolar inhibitor of farnesyl diphosphate synthase. *Arch Biochem Biophys*, 373, 231–241.
3. Black, D.M., Delmas, P.D., Eastell, R. et al. (2007). Once-yearly zoledronic acid for treatment of postmenopausal osteoporosis. *N Engl J Med*, 356, 1809–1822.
4. Chen, T., Berenson, J., Vescio, R. et al. (2002). Pharmacokinetics and pharmacodynamics of zoledronic acid in cancer patients with bone metastases. *J Clin Pharmacol*, 42, 1228–1236.
5. Coxon, F.P., & Rogers, M.J. (2003). The Role of Prenylated Small GTP-Binding Proteins in the Regulation of Osteoclast Function. *Calcif Tissue Int*, 72, 80–84.
6. Coxon, F.P., Thompson, K., Roelofs, A.J., Ebetino, F.H., & Rogers, M.J. (2008). Visualizing mineral binding and uptake of bisphosphonate by osteoclasts and non-resorbing cells. *Bone*, 42, 848–860.
7. Das, H., Wang, L., Kamath, A., & Bukowski, J.F. (2001). Vgamma2Vdelta2 T-cell receptor-mediated recognition of aminobisphosphonates. *Blood*, 98, 1616–1618.
8. Delmas, P.D., Adami, S., Strugala, C. et al. (2006). Intravenous ibandronate injections in postmenopausal women with osteoporosis: one-year results from the dosing intravenous administration study. *Arthritis Rheum*, 54, 1838–1846.
9. Dunford, J.E., Thompson, K., Coxon, F.P. et al. (2001). Structure-activity relationships for inhibition of farnesyl diphosphate synthase in vitro and inhibition of bone resorption in vivo by nitrogen-containing bisphosphonates. *J Pharmacol Exp Ther*, 296, 235–242.
10. Espinosa, E., Belmant, C., Pont, F. et al. (2001). Chemical synthesis and biological activity of bromohydrin pyrophosphate, a potent stimulator of human gamma delta T cells. *J Biol Chem*, 276, 18337–18344.
11. Frith, J.C., Monkkonen, J., Auriola, S., Monkkonen, H., & Rogers, M.J. (2001). The molecular mechanism of action of the anti-resorptive and anti-inflammatory drug clodronate: evidence for the formation in vivo of a metabolite that inhibits bone resorption and causes osteoclast and macrophage apoptosis. *Arth Rheum*, 44, 2201–2210.

12. Gober, H.J., Kistowska, M., Angman, L. et al. (2003). Human T cell receptor gammadelta cells recognize endogenous mevalonate metabolites in tumor cells. *J Exp Med*, 197, 163–168.
13. Hayday, A.C. (2000). [gamma][delta] cells: a right time and a right place for a conserved third way of protection. *Ann Rev Immunol*, 18, 975–1026.
14. Hewitt, R.E., Lissina, A., Green, A.E. et al. (2005). The bisphosphonate acute phase response: rapid and copious production of proinflammatory cytokines by peripheral blood gd T cells in response to aminobisphosphonates is inhibited by statins. *Clin Exp Immunol*, 139, 101–111.
15. Kunzmann, V., Bauer, E., Feurle, J. et al. (2000). Stimulation of gammadelta T cells by aminobisphosphonates and induction of antiplasma cell activity in multiple myeloma. *Blood*, 96, 384–392.
16. Kunzmann, V., Bauer, E., & Wilhelm, M. (1999). Gamma/delta T-cell stimulation by pamidronate. *N Engl J Med*, 340, 737–738.
17. Kunzmann, V., & Wilhelm, M. (2005). Anti-lymphoma effect of gammadelta T cells. *Leuk Lymphoma*, 46, 671–680.
18. Li, L., & Wu, C.Y. (2008). CD4+ CD25+ Treg cells inhibit human memory gammadelta T cells to produce IFN-gamma in response to M tuberculosis antigen ESAT-6. *Blood*, 111, 5629–5636.
19. Mariani, S., Muraro, M., Pantaleoni, F. et al. (2005). Effector gammadelta T cells and tumor cells as immune targets of zoledronic acid in multiple myeloma. *Leukemia*, 19, 664–670.
20. Miyagawa, F., Tanaka, Y., Yamashita, S., & Minato, N. (2001). Essential requirement of antigen presentation by monocyte lineage cells for the activation of primary human gamma delta T cells by aminobisphosphonate antigen. *J Immunol*, 166, 5508–5514.
21. Monkkonen, H., Auriola, S., Lehenkari, P. et al. (2006). A new endogenous ATP analog (ApppI) inhibits the mitochondrial adenine nucleotide translocase (ANT) and is responsible for the apoptosis induced by nitrogen-containing bisphosphonates. *Br J Pharmacol*, 147, 437–445.
22. Roelofs, A.J., Jauhiainen, M., Monkkonen, H. et al. (2009). Peripheral blood monocytes are responsible for gamma,delta T cell activation induced by zoledronic acid through accumulation of IPP/DMAPP. *Br J Haematol*, 144, 245–250.
23. Sato, M., Grasser, W., Endo, N. et al. (1991). Bisphosphonate action. Alendronate localization in rat bone and effects on osteoclast ultrastructure. *J Clin Invest*, 88, 2095–2105.
24. Sauty, A., Pecherstorfer, M., Zimmer-Roth, I. et al. (1996). Interleukin-6 and tumor necrosis factor alpha levels after bisphosphonates treatment in vitro and in patients with malignancy. *Bone*, 18, 133–139.
25. Schweitzer, D.H., Oostendorp-van de Ruit, M., Van der Pluijm, G., Lowik, C.W., & Papapoulos, S.E. (1995). Interleukin-6 and the acute phase response during treatment of patients with Paget's disease with the nitrogen-containing bisphosphonate dimethylaminohydroxypropylidene bisphosphonate. *J Bone Miner Res*, 10, 956–962.
26. Selander, K., Lehenkari, P., & Vaananen, H.K. (1994). The effects of bisphosphonates on the resorption cycle of isolated osteoclasts. *Calcif Tiss Int*, 55, 368–375.
27. Sicard, H., Ingoure, S., Luciani, B. et al. (2005). In vivo immunomanipulation of V gamma 9 V delta 2 T cells with a synthetic phosphoantigen in a preclinical nonhuman primate model. *J Immunol*, 175, 5471–5480.
28. Tanaka, Y., Morita, C.T., Tanaka, Y. et al. (1995). Natural and synthetic non-peptide antigens recognized by human gamma delta T cells. *Nature*, 375, 155–158.
29. Thiebaud, D., Sauty, A., Burckhardt, P. et al. (1997). An in vitro and in vivo study of cytokines in the acute-phase response associated with bisphosphonates. *Calcif Tissue Int*, 61, 386–392.
30. Thompson, K., Rogers, M.J., Coxon, F.P., & Crockett, J.C. (2006). Cytosolic entry of bisphosphonate drugs requires acidification of vesicles after fluid-phase endocytosis. *Mol Pharmacol*, 69, 1624–1632.
31. Thompson, K., & Rogers, M.J. (2004). Statins prevent bisphosphonate-induced gamma,delta-T-cell proliferation and activation in vitro. *J Bone Miner Res*, 19, 278–288.

32. van Beek, E., Pieterman, E., Cohen, L., Lowik, C., & Papapoulos, S. (1999). Nitrogen-containing bisphosphonates inhibit isopentenyl pyrophosphate isomerase/farnesyl pyrophosphate synthase activity with relative potencies corresponding to their antiresorptive potencies in vitro and in vivo. *Biochem Biophys Res Commun*, 255, 491–494.
33. van Beek, E., Pieterman, E., Cohen, L., Lowik, C., & Papapoulos, S. (1999). Farnesyl pyrophosphate synthase is the molecular target of nitrogen-containing bisphosphonates. *Biochem Biophys Res Commun*, 264, 108–111.
34. Wilhelm, M., Kunzmann, V., Eckstein, S. et al. (2003). Gammadelta T cells for immune therapy of patients with lymphoid malignancies. *Blood*, 102, 200–206.
35. Zhang, F.L., & Casey, P.J. (1996). Protein prenylation: molecular mechanisms and functional consequences. *Ann Rev Biochem*, 65, 241–269.
36. Zimolo, Z., Wesolowski, G., & Rodan, G.A. (1995). Acid extrusion is induced by osteoclast attachment to bone. Inhibition by alendronate and calcitonin. *J Clin Invest*, 96, 2277–2283.

Identification of Cell Cycle-Arrested Quiescent Osteoclast Precursors In Vivo

Naoyuki Takahashi, Akinori Muto, Atsushi Arai, and Toshihide Mizoguchi

Abstract How are sites suitable for osteoclastogenesis determined? We addressed this issue using in vivo and in vitro experimental systems. We first examined the formation of osteoclasts in ectopic bone induced by BMP-2. When collagen disks which contained BMP-2 (BMP-2-disks) or vehicle (control-disks) were implanted into wild-type mice, osteoclasts and osteoblasts appeared in the BMP-2-disks, but not in the control disks. RANKL-deficient (RANKL$^{-/-}$) mice exhibited osteopetrosis, with an absence of osteoclasts. BMP-2 and control disks were implanted into RANKL$^{-/-}$ mice, which were intraperitoneally injected with RANKL. Osteoclasts formed in the BMP-2-disks, but not in the control disks. In the BMP-2-disks, osteoclasts were observed in the vicinity of osteoblasts. Cell cycle-arrested quiescent osteoclast precursors (QOP) were identified as the committed osteoclast precursors in vitro. Experiments in vivo showed that QOPs survived for several weeks, and differentiated into osteoclasts in response to M-CSF and RANKL. QOPs were identified as RANK and c-Fms double-positive cells, and detected along bone surfaces in the vicinity of osteoblasts in RANKL$^{-/-}$ mice. QOPs were also observed in the ectopic bone induced by BMP-2 implanted into RANKL$^{-/-}$ mice, suggesting that QOPs were circulating. These results imply that osteoblasts support the homing of QOPs to bone tissues. In response to bone-resorbing stimuli, QOPs promptly differentiate into osteoclasts. Therefore, the distribution of QOPs appears to determine the correct site of osteoclastic development.

Keywords Osteoclast precursor · Osteoclastogenesis · Osteoblast · Osteoclast miche

Abbreviations

M-CSF	macrophage colony-stimulating factor

N. Takahashi (✉)
Division of Hard Tissue Research, Institute for Oral Science, Matsumoto Dental University, 1780 Gobara, Hiro-oka, Shiojiri-shi, Nagano 399-0781, Japan
e-mail: takahashinao@po.mdu.ac.jp

Y. Choi (ed.), *Osteoimmunology*, Advances in Experimental Medicine and Biology 658, DOI 10.1007/978-1-4419-1050-9_3, © Springer Science+Business Media, LLC 2010

RANK	receptor activator of NF κB
RANKL	receptor activator of NF κB ligand
PTH	parathyroid hormone
$1\alpha,25(OH)_2D_3$	1α,25-dihydroxyvitamin D_3
OPG	osteoprotegerin
WT	wild-type
Cdk	cyclin-dependent kinase
QOP	cell cycle-arrested quiescent osteoclast precursor
BMP-2	bone morphogenetic protein 2
TRAP	tartrate-resistant acid phosphatise
ALP	alkaline phosphatise
BrdU	bromodeoxyuridine
HSC	hematopoietic stem cell

1 Introduction

A series of experiments using a mouse-coculture system have established that osteoblasts are crucial to osteoclastic development [12, 15]. Osteoblasts express two cytokines essential for osteoclastogenesis: macrophage colony-stimulating factor (M-CSF) and receptor activator of NF κB ligand (RANKL) [1, 17]. RANKL is expressed as a membrane-associated cytokine in osteoblasts, in response to many bone-resorbing factors [13]. Mice deficient in M-CSF [3] or RANKL [5] develop osteopetrosis, with a complete lack of osteoclasts. M-CSF is constitutively expressed by osteoblasts, whereas the expression of RANKL is upregulated by osteotropic factors such as parathyroid hormone (PTH) and 1α,25-dihydroxyvitamin D_3 [$1\alpha,25(OH)_2D_3$]. Precursors of osteoclasts bind to RANKL expressed by osteoblasts through cell-cell interaction, and differentiate into osteoclasts in the presence of M-CSF [13]. Osteoblasts also produce osteoprotegerin (OPG), a soluble decoy receptor for RANKL [11, 18]. OPG inhibits osteoclastic differentiation and function by interrupting the interaction between RANKL and RANK. Thus, osteoblasts play a central role in osteoclastic bone resorption.

How are sites suitable for osteoclastogenesis determined? Several possibilities have been proposed. We believe that the site of RANKL and M-CSF expression determine where osteoclasts form, because the direct cell-cell interaction between osteoblasts and hematopoietic cells was required for osteoclasts to develop in a co-culture system [4, 14]. However, osteoclasts were generated in bone tissues in response to an injection of RANKL in RANKL-deficient ($RANKL^{-/-}$) mice, as if they were observed in wild-type (WT) mice. Osteoclasts were not observed in the soft tissues around the bone in the RANKL-treated mice. Similarly, when M-CSF was injected into M-CSF-deficient op/op mice, osteoclasts formed at suitable sites in bone. These results suggest that the presence of neither RANKL nor M-CSF is involved in determining the correct site for osteoclastogenesis.

Proliferation and differentiation are coordinated during the development of specialized cells [7, 10]. Cell proliferation is driven by heterodimeric kinases composed of a cyclin, a regulatory subunit, and a cyclin-dependent kinase (Cdk), a catalytic subunit. Cdk inhibitors such as the Cip/Kip family ($p21^{CIP1}$, $p27^{KIP1}$ and $p57^{KIP2}$) regulate the activity of Cdks. We examined the role of cell-cycle regulatory molecules in osteoclastogenesis using mouse bone marrow macrophage (BMMϕ) cultures treated with M-CSF and RANKL, and identified "cell cycle-arrested quiescent osteoclast precursors (QOPs)" as the direct osteoclast precursors in vitro. Studies in vivo showed that QOPs existed along bone tissues.

In this review, we propose that osteoblasts play important roles in osteoclastogenesis by providing a suitable microenvironment, the "osteoclast niche." In the osteoclast niche, osteoblasts support the homing of QOPs to bone tissues. In response to bone-resorbing stimuli, QOPs promptly differentiate into osteoclasts without cell-cycle progression. The distribution of QOPs or osteoclast niches appears to determine the site of osteoclastic development.

2 Formation of Osteoclasts in BMP-2-Induced Ectopic Bone

We first examined osteoclastogenesis in ectopic bone generated with bone morphogenetic protein 2 (BMP-2). Collagen disks which contained BMP-2 or vehicle were implanted into the left dorsal muscle pouches of WT mice [19]. The implants were recovered after 1 or 2 weeks. Tartrate-resistant acid phosphatase (TRAP, a marker of osteoclasts)-positive ($TRAP^+$) osteoclasts and alkaline phosphatase (ALP, a marker of osteoblasts)-positive (ALP^+) osteoblasts simultaneously appeared in BMP-2-disks but not in the control disks after implantation for 1 week. F4/80 (a marker of macrophages)-positive ($F4/80^+$) osteoclast precursors were similarly distributed in both BMP-2- and control disks. $TRAP^+$ cells were detected primarily in and around the layers of ALP^+ cells.

$TRAP^+$ osteoclasts were totally absent in tibiae of $RANKL^{-/-}$ mice [5]. Collagen disks which contained BMP-2 or vehicle were implanted into $RANKL^{-/-}$ mice, which were injected with RANKL for 8 days [19]. Many $TRAP^+$ osteoclasts were present in the tibiae, as in WT mice. ALP^+ cells appeared in BMP-2-disks implanted into $RANKL^{-/-}$ mice, but $TRAP^+$ cells did not (Fig. 1). Many $TRAP^+$ cells were also detected in the BMP-2-disks implanted into $RANKL^{-/-}$ mice injected with RANKL, whereas only a few $TRAP^+$ cells were observed in the control disks. Most $TRAP^+$ cells were in close proximity to ALP^+ cells. These results suggest that osteoblasts play critical roles in osteoclastogenesis, even when they do not express RANKL.

3 Identification of QOPs as the Direct Osteoclast Precursors

We next tried to identify the direct osteoclast precursor in vitro. The relationship between proliferation and differentiation of osteoclast precursors was examined in the first experiment [6]. BMMϕ (osteoclast precursors) differentiated into $TRAP^+$ multinucleated cells (osteoclasts) within 3 days in the presence of M-CSF and

(A) Experimental procedure

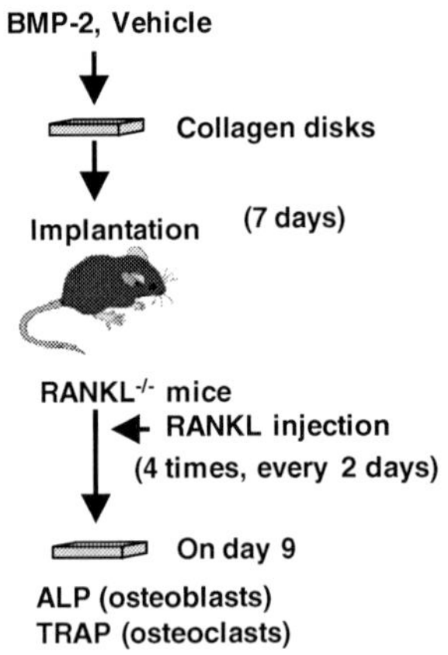

(B) Results

Implanted disks	BMP-2	BMP-2	Vehicle
RANKL injection	No	Yes	Yes
Appearance of osteoblasts (ALP-positive cells)	+	+	–
Appearance of osteoclasts (TRAP-positive cells)	–	+	–

Fig. 1 Osteoclastic development in BMP-2-generated ectopic bone in RANKL$^{-/-}$ mice. (**a**) Experimental procedure. Collagen disks containing BMP-2 or vehicle were implanted into the left dorsal muscle pouches of RANKL$^{-/-}$ mice. After implantation for 1 week, RANKL$^{-/-}$ mice were injected with RANKL (4 times every 2 days). The implants and tibiae were recovered on day nine, and subjected to TRAP and ALP staining. (**b**) BMP-2-gernerated ectopic bone. Appearance of TRAP$^+$ cell and ALP$^+$ cells in the implants is expressed as +/–

RANKL. The growth of BMMϕ treated with RANKL and M-CSF was retarded on days two and three, in comparison with that with M-CSF alone. Expression of cell-cycle regulatory molecules such as cyclins D1, D2, D3 and E1, and Cdks 2, 4 and 6 in BMMϕ was decreased by the treatment with RANKL. Purified osteoclasts failed to express cyclins D1, D2, D3 and E1, and Cdks 2 and 4. In contrast, expression of p27^{KIP1} was up-regulated in those cells. These results suggest that cell-cycle arrest in mature osteoclasts is maintained by the disappearance of cyclins and Cdks, and that p27^{KIP1} is involved in RANKL-induced osteoclastic differentiation of BMMϕ (Fig. 2). When hydroxyurea, an inhibitor of DNA replication, was added to the culture on day zero, RANKL-induced osteoclastic development was completely inhibited. In contrast, the formation of osteoclasts was accelerated by hydroxyurea, when it was added on day one. Hydroxyurea added on day two had no effect. Bromodeoxyuridine (BrdU) is a nucleoside analog that can be incorporated into dividing nuclei. When BrdU was added to BMMϕ cultures, together with RANKL on day zero, all the nuclei in RANKL-treated multinucleated cells were labeled with BrdU. When BrdU was added on day one (one day after treatment with RANKL), only 20% of the nuclei of multinucleated cells incorporated BrdU. Ki67 is a marker of actively cycling cells. Most nuclei of BMMϕ were positively stained with an anti-Ki67 antibody. In contrast, all of the nuclei in osteoclasts were negative for Ki67 (Ki67$^-$). These results suggest that cell-cycle progression and subsequent withdrawal in osteoclast precursors are required for their differentiation into the direct osteoclast precursors in vitro [6]. The direct osteoclast precursors were named "cell cycle-arrested quiescent osteoclast precursors (QOP)" (Fig. 2).

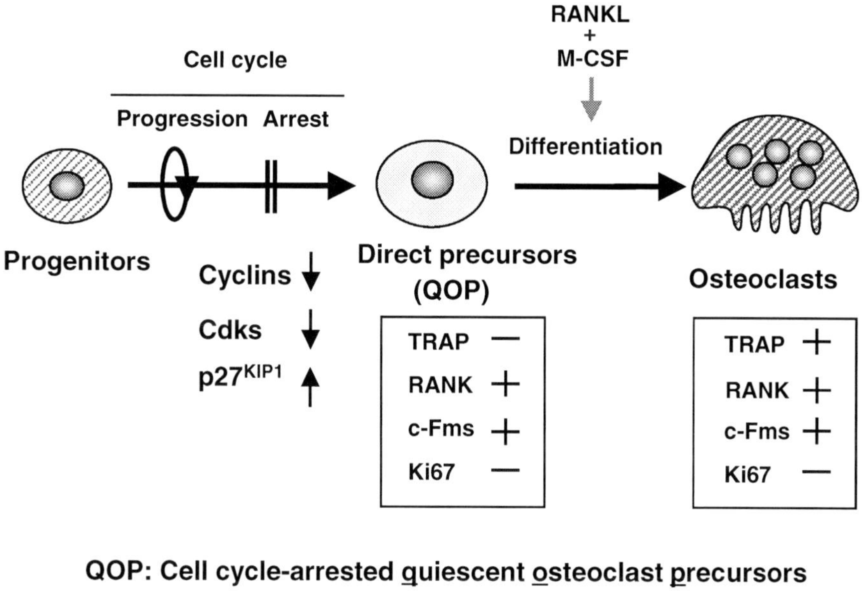

Fig. 2 Two cell cycle-related events in osteoclastogenesis. Cell cycle progression and subsequent withdrawal in osteoclast progenitors are required for their differentiation into osteoclasts in vitro. The direct osteoclast precursors are named "cell cycle-arrested quiescent osteoclast precursors (QOP)". Cell cycle arrest in QOPs is induced by the disappearance of cyclins and Cdks. $p27^{KIP1}$ is involved in the RANKL-induced differentiation of osteoclast progenitors into QOPs. Osteoclasts express TRAP, RANK, and c-Fms but not Ki67, while QOPs are believed to express RANK and c-Fms but not TRAP or Ki67

The definition of QOP is as follows: QOPs are cells that differentiate into osteoclasts without cell-cycle progression.

4 Characteristics of QOP

QOPs are believed to express both c-Fms (M-CSF receptor) and RANK (RANKL receptor) (Fig. 2). Using antibodies against c-Fms and RANK, mononuclear cells expressing both c-Fms and RANK ($Fms^+/RANK^+$ cells) were isolated from the bone marrow cell preparation of WT mice, and their characteristics were examined in more detail (Fig. 3). $Fms^+/RANK^+$ cells differentiated into osteoclasts in response to RANKL and M-CSF, even in the presence of hydroxyurea. Immunohistochemical staining showed that most of the $Fms^+/RANK^+$ cells were $F4/80^-$. The phagocytic activity of $Fms^+/RANK^+$ cells and their ability to differentiate into dendritic cells were much lower than those of BMMϕ. In contrast to BMMϕ, $Fms^+/RANK^+$ cells failed to proliferate in response to M-CSF. These results suggest that $Fms^+/RANK^+$ cells (QOP) isolated from bone marrow are

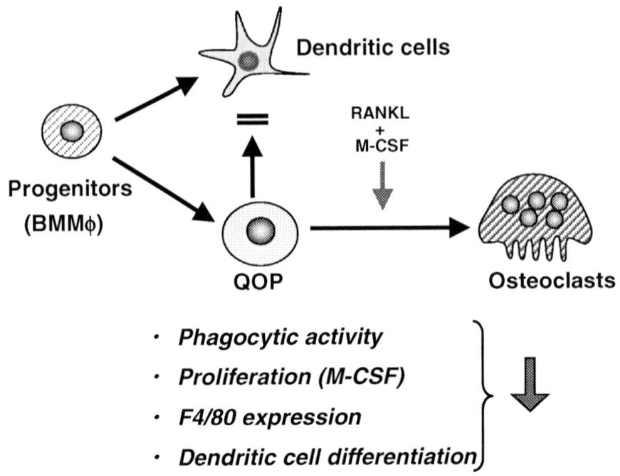

Fig. 3 Characteristics of QOPs. Characteristics of Fms$^+$/RANK$^+$ cells are compared with those of BMMϕ. Fms$^+$/RANK$^+$ cells have less phagocytic activity, proliferating activity in response to M-CSF, and potential to differentiate into dendritic cells than do BMMϕ. Most Fms$^+$/RANK$^+$ cells are negative for F4/80. QOPs appear to be committed precursors in an osteoclast lineage

committed precursors in an osteoclast lineage. We previously reported that postmitotic osteoclast precursors were formed in cocultures of osteoblasts and bone marrow cells [16]. Postmitotic osteoclast precursors differentiated into osteoclasts even in the presence of hydroxyurea, indicating that they are QOPs. These results suggest that osteoblasts play a role in the differentiation of hematopoietic progenitors into QOPs.

5 In Vivo Identification of QOPs

We next tried to identify QOPs in vivo [6]. We first examined in vivo labeling of the nuclei of osteoclasts with BrdU. BrdU in drinking water was administered to 7-week-old mice for 1 week, and the incorporation of BrdU into the nuclei of osteoclasts was evaluated in tibiae. Most nuclei of osteoclasts were BrdU-negative (BrdU$^-$) in 8-week-old mice. When 7-week-old mice were given BrdU for an additional 7 weeks, most of the bone marrow cells around osteoclasts were BrdU-positive (BrdU$^+$), but only 50% of the nuclei of osteoclasts were labeled with BrdU. Previous studies have shown that the lifespan of osteoclasts is 2–4 weeks in both humans and mice [9]. Therefore, these results suggest that the lifespan of QOPs is longer than 4 weeks.

We next examined whether osteoclasts are formed from cell cycle-arrested QOPs in response to several stimuli [6]. Osteoclasts are totally absent in RANKL$^{-/-}$ mice. RANKL$^{-/-}$ mice were given BrdU, and injected with RANKL for 2 days. Tibial sections were prepared and stained for TRAP and BrdU. Many osteoclasts were produced in response to RANKL injection. Interestingly, more than 70% of nuclei in osteoclasts were $BrdU^{-}$. These results suggest that RANKL stimulates differentiation of QOPs into osteoclasts in vivo. The effect of injecting M-CSF on BrdU labeling in osteoclasts was examined in M-CSF-deficient op/op mice. Op/op mice were given BrdU, and then injected with M-CSF for 7 days. $TRAP^{+}$ osteoclasts were formed in bone tissues in op/op mice in response to M-CSF. Again, more than 80% of nuclei in osteoclasts were BrdU-negative. These results also suggest that M-CSF, as well as RANKL, is not involved in the appearance of QOPs in mice. We further examined whether the differentiation of QOPs into osteoclasts is linked to the regulation of calcium metabolism in WT mice. Feeding WT mice a low-calcium diet induced osteoclastic bone resorption. The number of osteoclasts in tibiae was significantly increased in WT mice fed a low-calcium diet. However, the percentage of $BrdU^{+}$ nuclei in osteoclasts remained unchanged. These results suggest that osteoclasts produced by a low-calcium diet are formed from QOPs, and that all stimuli which induce osteoclastic bone resorption in vivo stimulate the differentiation of QOPs into osteoclasts, but not the differentiation of hematopoietic precursors into QOPs (Fig. 4).

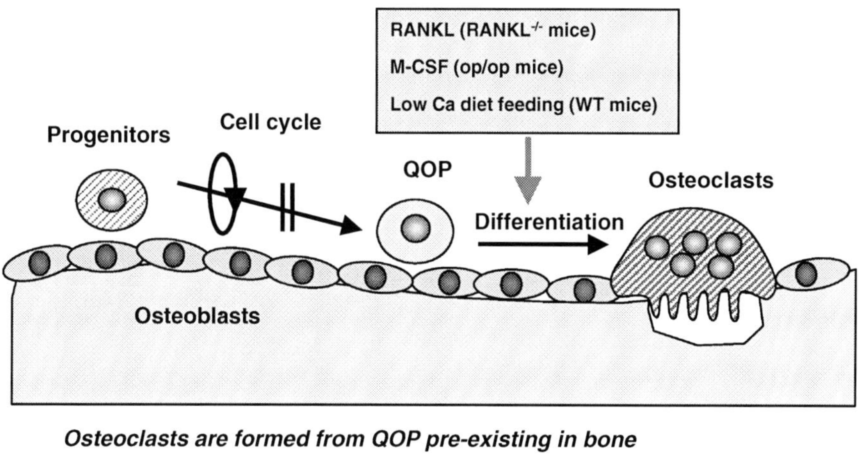

Fig. 4 Analysis of QOPs in adult mice. BrdU labeling shows that QOPs have quite a long lifespan. All stimuli, which induce osteoclastic bone resorption in vivo, stimulate differentiation of QOPs into osteoclasts, but not differentiation of osteoclast progenitors into QOPs

As shown in Fig. 2, QOPs are believed to express RANK and c-Fms but not TRAP or Ki67. Using these markers, the distribution of QOPs was examined in $RANKL^{-/-}$ mice, because QOPs but not osteoclasts exist in those mice. $Fms^{+}/RANK^{+}$ cells were detected along the surface of trabecular bones in $RANKL^{-/-}$ mice. The $RANK^{+}$ mononuclear cells were $Ki67^{-}$. These results suggest that QOPs are localized along bone surfaces in $RANKL^{-/-}$ mice. Then the distribution of QOPs and osteoblasts was compared in $RANKL^{-/-}$ mice, using anti-ALP antibodies. ALP^{+} cells were distributed along the surface of trabecular bone. The distribution of $RANK^{+}$ cells was quite similar to that of ALP^{+} cells in the trabecular bone area. $RANK^{+}$ cells were always observed in the vicinity of ALP^{+} osteoblasts.

These results suggest that osteoblasts are involved in the maintenance of QOPs in bone.

6 Osteoclast Niche Prepared by Osteoblasts

How do QOPs recognize and settle at sites suitable for osteoclastogenesis? We finally tried to identify QOPs in ectopic bone induced by BMP in RANKL-deficient mice [8]. BMP-2-disks or control disks were implanted into RANKL$^{-/-}$ mice for 2 weeks. The disks were then recovered and subjected to RANK and ALP staining. Bone had formed in the BMP-2-disks. ALP^{+} osteoblasts were observed along bone surfaces. ALP^{+} cells were not detected in the control disks. RANK^{+} cells were also detected along the surface of BMP-induced bone. The number of RANK^{+} cells was much higher in the BMP-2-disks than in the control disks. These results suggest that QOPs are circulating cells of hematopoietic origin, and that osteoblasts play a role in the homing of QOPs to bone tissues (Fig. 5). These homing and maintenance mechanisms occur in "osteoclast niches." Recent studies have shown that cells of the osteoblastic lineage function as a key component of the hematopoietic stem cell (HSC) niche, controlling the number of HSCs [2, 20]. Osteoblasts may be involved in maintaining QOPs as well as HSCs for long periods in a quiescent state. Unlike

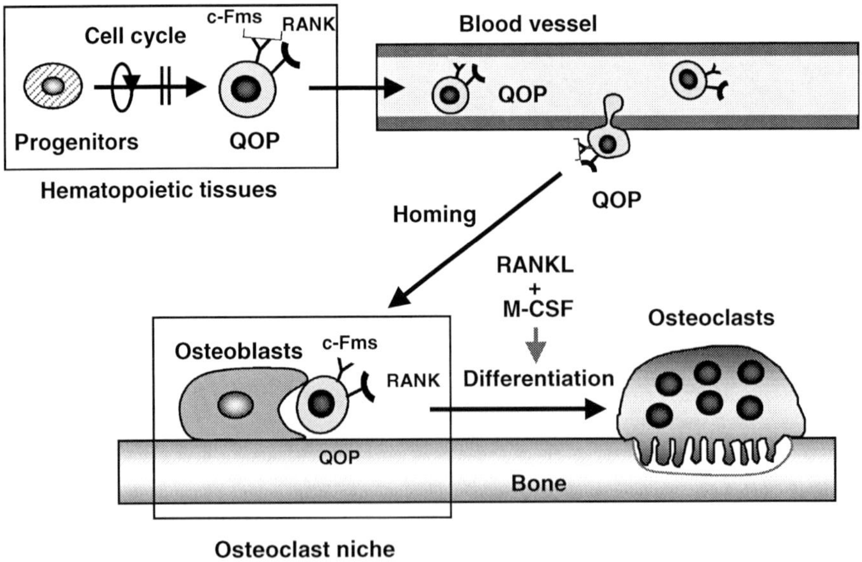

Fig. 5 Osteoclast niches prepared by osteoblasts. QOPs are circulating cells of hematopoietic origin. Osteoblasts may play a role in the homing of QOPs to bone tissues. The homing and maintenance of QOPs occur in "osteoclast niches". QOPs in the osteoclast niche promptly differentiate into osteoclasts in response to bone-resorption-inducing stimuli. The distribution of osteoclast niches appears to determine sites suitable for osteoclastogenesis

HSCs, QOPs are cells with transient characteristics, but lacking self-renewal capacity. The role of osteoblasts in an osteoclast niche may be different from that in the HSC niche. Therefore, we propose that the distribution of osteoclast niches determines the correct site of osteoclastic development. Further studies will elucidate the molecular mechanisms of the maintenance of QOPs in osteoclast niches.

References

1. Boyle, W.J., Simonet, W.S., Lacey, D.L. (2003). Osteoclast differentiation and activation. *Nature*, 423, 337–342.
2. Calvi, L.M., Adams, G.B., Weibrecht, K.W. et al. (2003). Osteoblastic cells regulate the haematopoietic stem cell niche. *Nature*, 425, 841–846.
3. Felix, R., Cecchini, M.G., & Fleisch, H. (1990). Macrophage colony stimulating factor restores in vivo bone resorption in the op/op osteopetrotic mouse. *Endocrinology*, 127, 2592–2594.
4. Itoh, K., Udagawa, N., Matsuzaki, K. et al. (2000). Importance of membrane- or matrix-associated forms of M-CSF and RANKL/ODF in osteoclastogenesis supported by SaOS-4/3 cells expressing recombinant PTH/PTHrP Receptors. *J Bone Miner Res*, 15, 1766–1775.
5. Kong, Y.Y., Feige, U., Sarosi, I. et al. (1999). Activated T cells regulate bone loss and joint destruction in adjuvant arthritis through osteoprotegerin ligand. *Nature*, 402, 304–309.
6. Mizoguchi, T., Muto, A., Udagawa, N. et al. (2009). Identification of cell cycle-arrested quiescent osteoclast precursors in vivo. *J Cell Biol*, 184, 541–554.
7. Morgan, D.O. (1995). Principles of CDK regulation. *Nature*, 374, 131–134.
8. Muto, A., Mizoguchi, T., Kobayashi, Y. et al. (2008). Cell cycle-arrested quiescent osteoclast precursors (QOP) circulate to settle in the osteoclast niche. *Aegean Conference Series*, 35, 95.
9. Riggs, B.L., & Parfitt, A.M. (2005). Drugs used to treat osteoporosis: the critical need for a uniform nomenclature based on their action on bone remodeling. *J Bone Miner Res*, 20, 177–184.
10. Sherr, C.J., Roberts, J.M. (1995). Inhibitors of mammalian G1 cyclin-dependent kinases. *Genes Dev*, 9, 1149–1163.
11. Simonet, W.S., Lacey, D.L., Dunstan, C.R. et al. (1997). Osteoprotegerin: a novel secreted protein involved in the regulation of bone density. *Cell*, 89, 309–319.
12. Suda, T., Takahashi, N., & Martin, T.J. (1992). Modulation of osteoclast differentiation. *Endocrine Rev*, 13, 66–80.
13. Suda, T., Takahashi, N., Udagawa, N. et al. (1999). Modulation of osteoclast differentiation and function by the new members of the tumor necrosis factor receptor and ligand families. *Endocr Rev*, 20, 345–357.
14. Takahashi, N., Akatsu, T., Udagawa, N. et al. (1988). Osteoblastic cells are involved in osteoclast formation. *Endocrinology*, 123, 2600–2602.
15. Takahashi, N., Udagawa, N., Takami, N. et al. (2002). Cells of bone: osteoclast generation. In: Raisz LG, Rodan GA, Bilezikian JP (eds.), Principles of bone biology. Academic Press, San Diego, pp. 109–126.
16. Takahashi, N., Udagawa, N., Tanaka, S. et al. (1994). Postmitotic osteoclast precursors are mononuclear cells which express macrophage-associated phenotypes. *Dev Biol*, 163, 212–221.
17. Takayanagi, H. (2007). Osteoimmunology: shared mechanisms and crosstalk between the immune and bone systems. *Nat Rev Immunol*, 7, 292–304.
18. Tsuda, E., Goto, M., Mochizuki, S. et al. (1997). Isolation of a novel cytokine from human fibroblasts that specifically inhibits osteoclastogenesis. *Biochem Biophys Res Commun*, 234, 137–142.

19. Yamamoto, Y., Udagawa, N., Matsuura, S. et al. (2006). Osteoblasts provide a suitable microenvironment for the action of receptor activator of NF-κB ligand. *Endocrinology*, 147, 3366–3374.
20. Zhang, J., Niu, C., Ye, L. et al. (2003). Identification of the haematopoietic stem cell niche and control of the niche size. *Nature*, 425, 836–841.

Megakaryocyte-Bone Cell Interactions

Melissa A. Kacena and Wendy A. Ciovacco

Abstract Emerging data show that megakaryocytes (MKs) play a role in the replication and development of bone cells. Both in vivo and in vitro evidence now show that MKs can have significant effects on cells of the osteoclast (OC) and osteoblast (OB) lineage, with obvious manifestations on bone phenotype, and probable significance for human pathology.

There are currently four mouse models in which increases in MK number lead to a specific bone pathology of markedly increased bone volume. While these models all achieve megakaryocytosis by different mechanisms, the resultant osteosclerotic phenotype observed is consistent across all models.

In vitro data suggest that MKs play a role in OC and OB proliferation and differentiation. While MKs express receptor activator of nuclear factor kappa B ligand (RANKL), a prerequisite for osteoclastogenesis, they also express many factors known to inhibit OC development, and co-cultures of MKs with OCs show a significant decrease in osteoclastogenesis. In contrast, MKs express several proteins with a known critical role in osteoblastogenesis and bone formation, and co-cultures of these two lineages result in up to a six-fold increase in OB proliferation and alterations in OB differentiation.

This research demonstrates the complex regulatory interactions at play between MKs and bone cells, and opens up potential targets for therapeutic intervention.

Keywords Megakaryocyte · Bone · Osteoblast · OsteoClast · TPO · NF-E2 · GATA-1 · Von Willenbrand

Abbreviations

BM	bone marrow
BMD	bone marrow density

M.A. Kacena (✉)
Department of Orthopaedic Surgery, Indiana University School of Medicine, 1120 South Drive, FH 115, Indianapolis, IN 46202
e-mail: mkacena@iupui.edu

Y. Choi (ed.), *Osteoimmunology*, Advances in Experimental Medicine and Biology 658, DOI 10.1007/978-1-4419-1050-9_4, © Springer Science+Business Media, LLC 2010

BMP	bone morphogenetic protein
BMSC	bone marrow stromal cells
GM-CSF	granulocyte/macrophage colony-stimulating factor
IFN-γ	interferon gamma
IL	interleukin
MK	megakaryocyte
OB	osteoblast
OC	osteoclast
OCIL	osteoclast inhibitory lectin
OPG	osteoprotegerin
PDGF	platelet-derived growth factor
Pt-vWD	platelet-type von Willenbrand disease
RANKL	receptor activator of nuclear factor kappa B ligand
TGF-β	transforming growth factor-β
TPO	thrombopoietin
TRAP	tartrate resistant acid phosphatase
VEGF	vascular endothelial growth factor

1 Introduction

While cells of the hematopoietic lineage and bone-cell progenitors (mesenchymal stem cells) both originate in the bone marrow (BM), and are thus intimately related in space, only in the last several decades has a functional link between the two lineages been sought. Hematopoietic and mesenchymal stem cells replicate in juxtaposition to each other, suggesting that local expression of signaling molecules by either lineage, or juxtacrine communication by cell contact, may have a direct or indirect effect on neighboring cells. In fact, research now shows that these two systems originally studied in isolation are, in fact, functionally connected, with the BM cavity not just housing hematopoietic progenitors, but interacting with these cells in a variety of ways.

MKs are platelet progenitor cells which primarily reside in the BM, and their reciprocal relationship with bone cells is one of the burgeoning areas of research just described. Multiple studies have now provided ample in vivo and in vitro evidence showing that MKs affect the development and differentiation of both OCs and OBs. This review attempts to summarize the current research involving MKs and bone-cell interactions, beginning by describing several mouse models with dysregulated megakaryopoiesis and resultant skeletal pathology, then looking at MK-OC interactions in vitro, and, finally, exploring the mounting evidence of the in vitro effect of MK on OB.

2 In Vivo Evidence: Mouse Models

There are currently four known mouse models with dysregulated megakaryopoiesis and resultant significant increases in bone volume. Mice that overexpress thrombopoietin (TPO), the main MK growth factor, have an approximate four-fold increase in BM MK number, and develop a concurrent osteosclerotic bone phenotype with increased bone mineral density (BMD) [1–5]. Mice with a deficiency in the transcription factors GATA-1 or NF-E2, necessary for terminal MK differentiation, develop marked increases in MK number with a concomitant reduction or total absence of platelet cells, and a tremendous increase in trabecular bone [6–8]. Most recently, a novel mouse model of platelet-type von Willenbrand disease (Pt-vWD) with a platelet phenotype identical to the human form of the disease, showed a marked increase in splenic MK with splenomegaly, and a high bone-mass phenotype, with decreased serum measures of bone resorption [9].

2.1 TPO Overexpressing Mice

Given that TPO is the major MK growth factor, and is essential for lineage proliferation and proper terminal differentiation, it is expected that a mouse model overexpressing TPO shows dysregulated MK reproduction and maturation. Perhaps most surprising is the dramatic phenotype manifested in these mice, as a myelofibrotic syndrome with osteosclerosis develops by nine months of age in mice repeatedly injected with TPO, or infected with a viral vector harboring the TPO gene [1–5].

Mice overexpressing TPO show a marked four-fold increase in absolute MK number, versus wild-type controls affecting all stages of differentiation [1, 2]. Additionally, levels of transforming growth factor-β (TGF-β) and platelet-derived growth factor (PDGF) are elevated two-fold and five-fold, respectively, versus controls [1–5]. TGF-β and PDGF are both growth factors expressed by MKs with proven marked effects on bone cells [10, 11]. This observed increase in MK expression of two growth factors, previously identified to stimulate OBs, raised the possibility that in this model MK secretion of TGF-β and PDGF resulted in the myelofibrotic, osteosclerotic phenotype seen in TPO overexpressing mice. This hypothesis was supported by studies illustrating that, in the absence of TGF-β, TPO overexpressing mice failed to exhibit the characteristic myelofibrotic, osteosclerotic syndrome. Irradiated, wild-type mice were engrafted with TGF-β–1–/– BM stem cells infected with a retrovirus-encoding murine TPO protein to induce TPO overexpression, but no myelofibrosis or osteosclerosis developed. Furthermore, when irradiated, wild-type mice were repopulated with wild-type TPO overexpressing stem cells. The femurs now showed significant myelofibrosis and osteosclerosis at 16-weeks post-transplantation [12]-clear evidence that the MK-secreted growth factor TFG-β plays a critical role in the final bone phenotype of TPO overexpressing mouse models.

Another study using transgenic mice constitutively expressing TPO showed predictably elevated numbers of MKs, with associated significant increases in plasma levels of both TGF-β-1 and osteoprotegerin (OPG) [13]. OPG inhibits osteoclastogenesis and is known to be expressed by MK [14–19]. This study implies that increased secretion of OPG, along with TGF-β, by MK contributes to the myelofibrotic, osteosclerotic phenotype of TPO overexpressing mice. Similarly, a study by Chagraoui et al. [20] implicates the upregulation of OPG and the associated inhibition of osteoclastogenesis in the pathogenesis of osteosclerosis. Here, irradiated wild-type or OPG –/– mice were repopulated with either wild-type or OPG –/– BM stem cells infected with a retrovirus-encoding murine TPO protein. While all mice showed subsequent increases in TGFβ-1 with associated myelofibrosis, only the wild-type recipients (engrafted with wild-type or OPG –/– stem cells) showed increased OPG plasma levels with associated osteosclerosis. As opposed to this dense bone phenotype, the OPG –/– recipients (engrafted with wild-type or OPG –/– stem cells) instead developed an osteroporotic phenotype [20]. These results suggest that the OPG secreted by the transplanted BM stromal cells and OB caused the osteosclerosis seen in the wild-type hosts.

Yet another investigation studied the direct effects of TPO on OC formation in vitro and demonstrated a TPO dose-dependent decrease in OC number [21]. This effect was most likely mediated by increased MK number, in response to increased TPO stimulation. This idea is further explored in Section 3. Thus, it appears that TPO plays an indirect role in bone turnover by its proliferative effect on MK.

In summation, TPO overexpressing mice exhibit marked increases in MK number with simultaneous increases in BMD mediated by various MK-secreted cytokines, specifically TFG-β and OPG.

2.2 GATA-1 and NF-E2 Deficient Mice

Pluripotential hematopoietic stem cells give rise to MKs through a stepwise differentiation process, with progression through each phase mediated by specific transcription factors, ultimately resulting in terminally differentiated MKs capable of releasing platelets. The selective loss of any of the transcription factors regulating MK differentiation results in arrested development and accumulation of cells at the latest stage of maturation. Specifically, the loss of either GATA-1 or NF-E2 MK transcription factors have been shown to produce dysregulated megakaryopoiesis, with GATA-1 knock-down, or NF-E2 knock-out mouse models both displaying marked megakaryocytosis and a paradoxical thrombocytopenia [6, 7].

GATA-1 transcription factor is one of the six members of the GATA family of zinc-finger transcription factors (GATA-1 through GATA-6 in vertebrates), with DNA-binding activity in the C-terminal zinc finger of GATA's single polypeptide chain. GATA-1, previously thought to be necessary exclusively for erythroid lineage development, is now known to play a critical role in MK differentiation. Its expression is restricted to the hematopoietic lineage almost entirely, with known expression by multi-potential hematopoietic progenitors, mast cells, and MKs [22]. In GATA-1 knock-down models, the MK number in the BM and spleen increases approximately ten-fold, while the peripheral platelet count is decreased to 15% of wild-type controls [23]. We have demonstrated that GATA-1 deficient mice have a higher bone mass than controls, displaying more than a three-fold increase in bone volume and bone formation detectable after ~4 months [8]. At ages over a year, these animals ultimately develop a myelofibrotic phenotype [24]. MKs from GATA-1 deficient mice are also less differentiated than wild-type controls, as measured by lower levels of expression of TFG-β, PDGF, and vascular endothelial growth factor (VEGF) [10].

NF-E2 transcription factor is a heterodimeric nuclear protein composed of two polypeptide chains, both belonging to the basic leucine zipper family of transcription factors. The p18 subunit is ubiquitously expressed [25, 26], while the hematopoietic-specific 45-kDa subunit's expression is restricted to erythroid precursors, MKs, mast cells, and multipotential progenitors, similar to GATA-1. Like the GATA-1 knock-down model, MK number in the BM and spleen of adult mice lacking p45 NF-E2 is increased, though not as profoundly (two-five-fold), and a severe thrombocytopenia develops due to maturational arrest of MK development, with essentially no detectable platelets in the peripheral blood (less than 5% of control levels) [6]. Interestingly, we have shown that NF-E2 deficient mice also develop a high bone-mass phenotype, with up to five-fold increases in bone volume and bone-formation parameters [8, 27]. NF-E2 deficient mice respond to exogenous TPO with a marked proliferation, but there is no detectable increase in platelet production in vivo; and although MK number is markedly elevated, TPO levels are normal in NF-E2 deficient mice [6, 28].

Mice deficient in GATA-1 and NF-E2 transcription factors necessary for proper MK differentiation have strikingly similar phenotypes, characterized by marked megakaryocytosis, thrombocytopenia, and significantly increased BMD.

2.3 Murine Model of Pt-vWD

Suva et al. [9]. recently developed a mouse model of Pt-vWD by creating a transgenic cassette containing the human Pt-vWD point mutation (G233V) capable of expression in a murine colony. This gain of function mutation affects the GP-Ibα subunit of the platelet glycoprotein Ib-IX receptor complex, which normally binds von Willebrand factor and promotes platelet adhesion during vascular damage. A limited number of mutations, such as G233V, are known to alter the glycoprotein receptor complex configuration, while still allowing interaction with soluble von Willebrand factor. This is the pathogenesis behind Pt-vWD [9].

The phenotype caused by the Pt-vWD mutation in the mice mirrored the human disorder, with platelet dysfunction and resultant impaired hemostasis. The G233V mutant mice had a modest thrombocytopenia, with platelet counts reduced by 20%, and showed significantly increased bleeding time versus wild-type mice expressing the normal human GP-Ibα subunit. Mutant mice were also found to have spleens that were 2.5 times as large as wild-type controls, leading investigators to histological evaluation and discovery of a tissue-specific megakaryocytosis, with a marked increase in splenic MK number. Interestingly, there was no increase in the number of MKs in the BM of the mutant mice. The observation of MK-dense spleens prompted an evaluation of the mouse BM to determine if MK dysregulation was affecting skeletal homeostasis. Histological examination of mutant mouse BM revealed a high bone-mass phenotype detectable as early as two months, and bone mass progressively increased with age. Mutant bones also exhibited increased biomechanical strength versus wild-type controls [9].

Further BM analysis revealed a decrease in OC number, with no matched decrease in OBs, suggesting the high bone-mass phenotype was attributable to decreased OC number and resultant decreased bone breakdown. Additionally, ex vivo cultures further demonstrated that BM cultures from the Pt-vDW mice showed a significant decrease in the number of tartrate-resistant acid phosphatase (TRAP)-positive OC-like cells, versus wild-type controls (TRAP is a biomarker of OC number). Ex vivo cultures failed to show a difference in OB number or differentiation versus wild-type controls [9].

The investigators attempted to identify expression of the GP Ib-IX receptor complex on OCs to determine if the Pt-vWD mutation was having a direct effect on OC proliferation, but transcript profiling and immunofluoresence failed to find GP-Ib-IX expression by OCs. This supports prior evidence that the receptor complex is expressed exclusively on platelets as an MK lineage-specific gene product [9].

The Pt-vWD murine model displays a phenotype similar to the human disease it was designed to mimic. In addition to disrupted hemostasis, these mice exhibit increases in splenic MK number, increased BMD, and a decreased number of OCs. Although the exact mechanism by which this tissue-specific megakaryocytosis contributes to decreased OC population and decreased osteoclastogenesis remains to be elucidated, this study further confirms the specific links between MK function, platelet development, and OC formation and differentiation [9].

Taken collectively, the four mouse models detailed above illustrate the complex role of MKs in regulating skeletal mass, and show that disruptions in various points of MK differentiation and development consistently lead to an osteosclerotic phenotype.

3 MK-OC Interactions

MKs have the ability to affect, directly and indirectly, osteoclastogenesis, as evidenced by recent studies. The direct effects of MKs on OC differentiation are complex because MKs express proteins that enhance osteoclastogenesis, and other proteins known to inhibit it. RANKL is critical for OC development, and while it is thought that OB and OB precursors account for the majority of RANKL found within the BM, MKs have also been shown to express RANKL [14, 19, 29, 30]. However, MKs can also inhibit osteoclastogenesis as studies have demonstrated that MKs express or secrete several factors known to downregulate OC terminal differentiation, specifically OPG (an antagonist of RANK signaling), interleukin (IL)-10, IL-13, TGF-β and granulocyte/macrophage colony-stimulating factor (GM-CSF) [11, 14–18, 31–34]. Thus, MKs have the potential to affect significantly the OC number through their expression of factors that both promote and retard osteoclastogenesis. Notably, our group, as well as others, has demonstrated that when OC progenitors are cultured with MKs, or in an MK-conditioned medium, in vitro OC development is significantly inhibited by up to ten-fold [21, 31, 35]. We have also demonstrated that OPG expression alone is not responsible for this inhibition, as MKs derived from OPG-deficient mice were also able to inhibit OC formation in vitro [31]. As a result, we are currently working to isolate and identify the MK-secreted OC inhibitory factor using biochemical separation techniques, including HPLC. We have identified a single fraction with strong inhibitory activity containing less than 30 proteins. Interestingly, none of the major factors known to inhibit osteoclastogenesis, namely OPG, IL-4, IL-10, IL-12, IL-13, IL-18, interferon gamma (IFN-γ), TGFβ, GM-CSF, OC inhibitory lectin (OCIL), calcitonin, amylin, and calcitonin gene-related peptide, were present in the isolated fraction [31]. Therefore, while direct, in vitro evidence of MK-induced inhibition of OC development exists, the factor(s) or mechanism(s) responsible remain unknown.

MKs can also indirectly influence osteoclastogenesis. MKs increase OB and fibroblast proliferation by direct cell-to-cell contact [8, 36–40]. OB and fibroblasts are cells with known effects on osteoclastogenesis, including the expression of RANKL and OPG [41–48]. In addition to increasing OB proliferation, MKs increase OB expression of OPG when co-cultured [14]. As described previously, OPG inhibits osteoclastogenesis, so a potential indirect path for MK inhibition of OC formation exists.

In conclusion, most direct and indirect evidence suggests that MKs act to inhibit the differentiation of cells of the OC lineage. MKs express numerous proteins known to inhibit OC progenitor proliferation; co-culturing MKs with OC progenitors results in significant inhibition of OC formation, and MKs indirectly increase OPG secretion by both OB and fibroblasts. However, under certain physiological circumstances, MK expression of RANKL may be an important stimulator of osteoclastogenesis, particularly during inflammatory responses such as rheumatoid arthritis [49].

4 MK-OB Interactions

Studies have shown that MKs affect OB development by the secretion of bone matrix proteins and growth factors, by affecting OB differentiation, and-as mentioned previously-by directly increasing OB proliferation.

MKs or their platelet products secrete multiple bone matrix proteins, namely, osteocalcin; osteonectin; bone sialoprotein, and osteopontin [50–54]. MKs also secrete multiple growth factors crucial for bone remodeling, including, TGFβ-1; PDGF; VEGF, and bone morphogenetic protein (BMP)–2, –4, and –6 [10, 11, 55]. Therefore, MKs could impact bone formation and bone remodeling, especially in the setting of elevated local concentrations.

Our in vitro evidence demonstrates that MKs enhance OB proliferation three-six-fold by a direct cell-to-cell contact mechanism [8]. Additionally, co-cultures of MKs with BM stromal cells (BMSC) also enhance osteoblastogenesis, again by a mechanism requiring direct cell-to-cell contact [36]. Although these data demonstrate that MKs mediate OB proliferation by a juxtacrine signaling mechanism, the exact method remains to be identified.

MKs appear to affect both OB proliferation and differentiation. With respect to differentiation, Bord et al. [14], demonstrated that MKs stimulated OB differentiation as defined by increased OB expression of collagen (COL1 A1) and OPG. In this study, OB mRNA was examined 1–2 days following co-culture [14]. We recently concluded long-term studies where MKs and OBs were co-cultured for 14 days and examined the effect on OB differentiation, using both mRNA and functional assays. In our studies we examined OB expression of type I collagen, osteocalcin, and alkaline phosphatase. Additionally, we performed two functional in vitro assays to measure alkaline phosphatase activity and mineralization (calcium deposition). Unexpectedly, we found that, despite inducing a substantial increase in OB proliferation, MKs inhibited all markers of OB differentiation [56]. These data were unexpected, in light of data presented by Bord et al. [14]; however, differences in findings could easily be accounted for by the differences in culture duration.

5 Conclusion

It is now eminently clear that the hematopoietic and bone lineages are connected not just by proximity, but by functionality, and that MK-induced effects on OBs and OCs are multiple and complex. Several mouse models demonstrate that alterations and increases in cells of the MK lineage can lead to increases in bone volume in vivo; and, in vitro data shows a primarily inhibitory effect of MKs on OCs, and substantial pleiotropic effects on OBs. While the exact mechanisms of inhibition and induction need to be further elucidated, this represents an obvious target for therapeutic intervention, with the possibility of significant clinical impact on various debilitating bone diseases such as osteoporosis, or more obscure diseases of dysregulated bone turnover such as Paget's disease. As we continue to unravel exactly how local concentration of MK number and secretory products contribute to bone metabolism, pharmacologic and even genetic therapies for important public health problems may be discovered.

Acknowledgments The authors wish to thank Dr. Amanda Taylor for her critical review of this study. This work was supported in part by NIH/NIAMS grant AR055269 (MAK).

References

1. Frey, B.M., Rafii, S., Crystal, R.G. et al. (1998). Adenovirus long-term expression of thrombopoietin in vivo: a new model for myeloproliferative syndrome and osteomyelofibrosis. *Schweiz Med Wochenschr*, 128, 1587–1592
2. Frey, B.M., Rafii, S., Teterson, M. et al. (1998). Adenovector-mediated expression of human thrombopoietin cDNA in immune-compromised mice: insights into the pathophysiology of osteomyelofibrosis. *J Immunol*, 160, 691–699

3. Yan, X.Q., Lacey, D., Fletcher, F. et al. (1995). Chronic exposure to retroviral vector encoded MGDF (mpl-ligand) induces lineage-specific growth and differentiation of megakaryocytes in mice. *Blood*, 86, 4025–4033
4. Yan, X.Q., Lacey, D., Hill, D. et al. (1996). A model of myelofibrosis and osteosclerosis in mice induced by overexpressing thrombopoietin (mpl ligand): reversal of disease by bone marrow transplantation. *Blood*, 88, 402–409
5. Villeval, J.L., Cohen-Solal, K., Tulliez, M. et al. (1997). High thrombopoietin production by hematopoietic cells induces a fatal myeloproliferative syndrome in mice. *Blood*, 90, 4369–4383
6. Shivdasani, R.A., Rosenblatt, M.F., Zucker-Franklin, D. et al. (1995). Transcription factor NF-E2 is required for platelet formation independent of the actions of thrombopoietin/MGDF in megakaryocyte development. *Cell*, 81, 695–704
7. Shivdasani, R.A., Fujiwara, Y., McDevitt, M.A. et al. (1997). A lineage-selective knockout establishes the critical role of transcription factor GATA-1 in megakaryocyte growth and platelet development. *EMBO J*, 16, 3965–3973
8. Kacena, M.A., Shivdasani, R.A., Wilson, K. et al. (2004). Megakaryocyte-osteoblast interaction revealed in mice deficient in transcription factors GATA-1 and NF-E2. *J Bone Miner Res*, 19, 652–660
9. Suva, L.J., Hartman, E., Dilley, J.D. et al. (2008). Platelet dysfunction and a high bone mass phenotype in a murine model of platelet-type von Willebrand disease. *Am J Pathol*, 172, 430–439
10. Vannucchi, A.M., Bianchi, L., Cellai, C. et al. (2002). Development of myelofibrosis in mice genetically impaired for GATA-1 expression (GATA-1(low) mice). *Blood*, 100, 1123–1132
11. Wickenhauser, C., Hillienhof, A., Jungheim, K. et al. (1995). Detection and quantification of transforming growth factor beta (TGF-beta) and platelet-derived growth factor (PDGF) release by normal human megakaryocytes. *Leukemia*, 9, 310–315
12. Chagraoui, H., Komura, E., Tulliez, M. et al. (2002). Prominent role of TGF-beta 1 in thrombopoietin-induced myelofibrosis in mice. *Blood*, 100, 3495–3503
13. Kakumitsu, H., Kamezaki, K., Shimoda, K. et al. (2005). Transgenic mice overexpressing murine thrombopoietin develop myelofibrosis and osteosclerosis. *Leuk Res*, 29, 761–769
14. Bord, S., Frith, E., Ireland, D.C. et al. (2005). Megakaryocytes modulate osteoblast synthesis of type-l collagen, osteoprotegerin, and RANKL. *Bone*, 36, 812–819
15. Bord, S., Frith, E., Ireland, D.C. et al. (2004). Synthesis of osteoprotegerin and RANKL by megakaryocytes is modulated by oestrogen. *Br J Haematol*, 126, 244–251
16. Bord, S., Ireland, D.C., Beavan, S.R. et al. (2003). The effects of estrogen on osteoprotegerin, RANKL, and estrogen receptor expression in human osteoblasts. *Bone*, 32, 136–141
17. Pearse, R.N., Sordillo, E.M., Yaccoby, S. et al. (2001). Multiple myeloma disrupts the TRANCE/osteoprotegerin cytokine axis to trigger bone destruction and promote tumor progression. *Proc Natl Acad Sci USA*, 98, 11581–11586
18. Chagraoui, H., Sabri, S., Capron, C. et al. (2003). Expression of osteoprotegerin mRNA and protein in murine megakaryocytes. *Exp Hematol*, 31, 1081–1088
19. Kartsogiannis, V., Zhou, H., Horwood, N.J. et al. (1999). Localization of RANKL (receptor activator of NF kappa B ligand) mRNA and protein in skeletal and extraskeletal tissues. *Bone*, 25, 525–534
20. Chagraoui, H., Tulliez, M., Smayra, T. et al. (2003). Stimulation of osteoprotegerin production is responsible for osteosclerosis in mice overexpressing TPO. *Blood*, 101, 2983–2989
21. Wakikawa, T., Shioi, A., Hino, M. et al. (1997). Thrombopoietin inhibits in vitro osteoclastogenesis from murine bone marrow cells. *Endocrinology*, 138, 4160–4166
22. Orkin, S.H. (1992). GATA-binding transcription factors in hematopoietic cells. *Blood*, 80, 575–581
23. Vyas, P., Ault, K., Jackson, C.W. et al. (1999). Consequences of GATA-1 deficiency in megakaryocytes and platelets. *Blood*, 93, 2867–2875

24. Vannucchi, A.M., Bianchi, L., Paoletti, F. et al. (2005). A pathobiologic pathway linking thrombopoietin, GATA-1, and TGF-beta1 in the development of myelofibrosis. *Blood*, 105, 3493–3501
25. Andrews, N.C., Erdjument-Bromage, H., Davidson, M.B. et al. (1993). Erythroid transcription factor NF-E2 is a haematopoietic-specific basic-leucine zipper protein. *Nature*, 362, 722–728
26. Andrews, N.C., Kotkow, K.J., Ney, P.A. et al. (1993). The ubiquitous subunit of erythroid transcription factor NF-E2 is a small basic-leucine zipper protein related to the v-maf oncogene. *Proc Natl Acad Sci U S A*, 90, 11488–11492
27. Kacena, M.A., Gundberg, C.M., Nelson, T. et al. (2005). Loss of the transcription factor p45 NF-E2 results in a developmental arrest of megakaryocyte differentiation and the onset of a high bone mass phenotype. *Bone*, 36, 215–223
28. Shivdasani, R.A., Fielder, P., Keller, G.A. et al. (1997). Regulation of the serum concentration of thrombopoietin in thrombocytopenic NF-E2 knockout mice. *Blood*, 90, 1821–1827
29. Kong, Y.Y., Yoshida, H., Sarosi, I. et al. (1999). OPGL is a key regulator of osteoclastogenesis, lymphocyte development and lymph-node organogenesis. *Nature*, 397, 315–323
30. Manabe, N., Kawaguchi, H., Chikuda, H. et al. (2001). Connection between B lymphocyte and osteoclast differentiation pathways. *J Immunol*, 167, 2625–2631
31. Kacena, M.A., Nelson, T., Clough, M.E. et al. (2006). Megakaryocyte-mediated inhibition of osteoclast development. *Bone*, 39, 991–999
32. Jiang, S., Levine, J.D., Fu, Y. et al. (1994). Cytokine production by primary bone marrow megakaryocytes. *Blood*, 84, 4151–4156
33. Soslau, G., Morgan, D.A., Jaffe, J.S. et al. (1997). Cytokine mRNA expression in human platelets and a megakaryocytic cell line and cytokine modulation of platelet function. *Cytokine*, 9, 405–411
34. Wickenhauser, C., Lorenzen, J., Thiele, J. et al. (1995). Secretion of cytokines (interleukins-1 alpha, -3, and -6 and granulocyte-macrophage colony-stimulating factor) by normal human bone marrow megakaryocytes. *Blood*, 85, 685–691
35. Beeton, C.A., Bord, S., Ireland, D. et al. (2006). Osteoclast formation and bone resorption are inhibited by megakaryocytes. *Bone*, 39, 985–990
36. Miao, D., Murant, S., Scutt, N. et al. (2004). Megakaryocyte-bone marrow stromal cell aggregates demonstrate increased colony formation and alkaline phosphatase expression in vitro. *Tissue Eng*, 10, 807–817
37. Schmitz, B., Thiele, J., Otto, F. et al. (1996). Interactions between endogeneous lectins and fucosylated oligosaccharides in megakaryocyte-dependent fibroblast growth of the normal bone marrow. *Leukemia*, 10, 1604–1614
38. Schmitz, B., Thiele, J., Witte, O. et al. (1995). Influence of cytokines (IL-1 alpha, IL-3, IL-11, GM-CSF) on megakaryocyte-fibroblast interactions in normal human bone marrow. *Eur J Haematol*, 55, 24–32
39. Schmitz, B., Wickenhauser, C., Thiele, J. et al. (1999). Megakaryocyte induced fibroblast proliferation is enhanced by costimulation with IL-6/IL-3 and dependent on secretory and adhesion events. *Leuk Res*, 23, 723–729
40. Schmitz, B., Thiele, J., Otto, F. et al. (1998). Evidence for integrin receptor involvement in megakaryocyte-fibroblast interaction: a possible pathomechanism for the evolution of myelofibrosis. *J Cell Physiol*, 176, 445–455
41. Gao, Y.H., Shinki, T., Yuasa, T. et al. (1998). Potential role of cbfa1, an essential transcriptional factor for osteoblast differentiation, in osteoclastogenesis: regulation of mRNA expression of osteoclast differentiation factor (ODF). *Biochem Biophys Res Commun*, 252, 697–702
42. Hofbauer, L.C., Dunstan, C.R., Spelsberg, T.C. et al. (1998). Osteoprotegerin production by human osteoblast lineage cells is stimulated by vitamin D, bone morphogenetic protein-2, and cytokines. *Biochem Biophys Res Commun*, 250, 776–781

43. Hofbauer, L.C., & Heufelder, A.E. (1998). Osteoprotegerin and its cognate ligand: a new paradigm of osteoclastogenesis. *Eur J Endocrinol*, 139, 152–154
44. Vidal, O.N., Sjogren, K., Eriksson, B.I. et al. (1998). Osteoprotegerin mRNA is increased by interleukin-1 alpha in the human osteosarcoma cell line MG-63 and in human osteoblast-like cells. *Biochem Biophys Res Commun*, 248, 696–700
45. Udagawa, N., Takahashi, N., Jimi, E. et al. (1999). Osteoblasts/stromal cells stimulate osteoclast activation through expression of osteoclast differentiation factor/RANKL but not macrophage colony-stimulating factor: receptor activator of NF-kappa B ligand. *Bone*, 25, 517–523
46. Nakano, K., Okada, Y., Saito, K. et al. (2004). Induction of RANKL expression and osteoclast maturation by the binding of fibroblast growth factor 2 to heparan sulfate proteoglycan on rheumatoid synovial fibroblasts. *Arthritis Rheum*, 50, 2450–2458
47. Mitani, M., Miura, Y., Saura, R. et al. (2005). Estrogen specifically stimulates expression and production of osteoprotegerin from rheumatoid synovial fibroblasts. *Int J Mol Med*, 15, 827–832
48. Wei, X., Zhang, X., Zuscik, M.J. et al. (2005). Fibroblasts express RANKL and support osteoclastogenesis in a COX-2-dependent manner after stimulation with titanium particles. *J Bone Miner Res*, 20, 1136–1148
49. Kacena, M.A., & Horowitz, M.C. (2006). The role of megakaryocytes in skeletal homeostasis and rheumatoid arthritis. *Curr Opin Rheumatol*, 18, 405–410
50. Thiede, M.A., Smock, S.L., Petersen, D.N. et al. (1994). Presence of messenger ribonucleic acid encoding osteocalcin, a marker of bone turnover, in bone marrow megakaryocytes and peripheral blood platelets. *Endocrinology*, 135, 929–937
51. Kelm, R.J., Jr., Hair, G.A., Mann, K.G. et al. (1992). Characterization of human osteoblast and megakaryocyte-derived osteonectin (SPARC). *Blood*, 80, 3112–3119
52. Breton-Gorius, J., Clezardin, P., Guichard, J. et al. (1992). Localization of platelet osteonectin at the internal face of the alpha-granule membranes in platelets and megakaryocytes. *Blood*, 79, 936–941
53. Chenu, C., & Delmas, P.D. (1992). Platelets contribute to circulating levels of bone sialoprotein in human. *J Bone Miner Res*, 7, 47–54
54. Frank, J.D., Balena, R., Masarachia, P. et al. (1993). The effects of three different demineralization agents on osteopontin localization in adult rat bone using immunohistochemistry. *Histochemistry*, 99, 295–301
55. Sipe, J.B., Zhang, J., Waits, C. et al. (2004). Localization of bone morphogenetic proteins (BMPs)-2, -4, and -6 within megakaryocytes and platelets. *Bone*, 35, 1316–1322
56. Ciovacco, W.A., Goldberg, C.G., Taylor, A.F. et al. (2009). The role of gap junctions in megakaryocyte-mediated osteoblast proliferation and differentiation. *Bone*, 44, 80–86

Regulation of Osteoblast Differentiation by Runx2

Toshihisa Komori

Abstract Runx2 protein is first detected in preosteoblasts, and the expression is upregulated in immature osteoblasts, but downregulated in mature osteoblasts. Runx2 is the first transcription factor required for determination of the osteoblast lineage, while Sp7 and canonical Wnt-signaling further direct the fate of mesenchymal cells to osteoblasts, blocking their differentiation into chondrocytes. Runx2 induces the differentiation of multipotent mesenchymal cells into immature osteoblasts, directing the formation of immature bone, but Runx2 inhibits osteoblast maturation and mature bone formation. Normally, the protein level of Runx2 in osteoblasts reduces during bone development, and osteoblasts acquire mature phenotypes, which are required for mature bone formation. Furthermore, Runx2 triggers the expression of major bone matrix genes during the early stages of osteoblast differentiation, but Runx2 is not essential for the maintenance of these gene expressions in mature osteoblasts.

Keywords Runx2 · Sp7 · Canonical Wnt Signaling · Osteoblast · Osteopontin · Osteocalcin

1 Introduction

Runx2 is a transcription factor that belongs to the Runx family, and contains the DNA binding domain runt, which is homologous to the *Drosophila* pair rule gene *runt* [12]. Runx2 has been shown to be an essential transcription factor for osteoblast differentiation. In studies, *Runx2*-deficient mice showed a complete lack of bone formation due to the absence of osteoblasts [13, 18]. Cbfβ, a cotranscription factor, is required for DNA binding of Runx2 and Rux2-dependent osteoblast differentiation and bone formation [21]. Runx2 forms heterodimers with Cbfβ and recognizes the consensus sequence, PyGPyGGTPy [12].

T. Komori (✉)
Unit of Basic Medical Sciences, Department of Cell Biology, Nagasaki University Graduate School of Biomedical Sciences, 1-7-1 Sakamoto, Nagasaki 852-8588, Japan
e-mail: komorit@nagasaki-u.ac.jp

Y. Choi (ed.), *Osteoimmunology*, Advances in Experimental Medicine and Biology 658, DOI 10.1007/978-1-4419-1050-9_5, © Springer Science+Business Media, LLC 2010

2 Expression of Runx2 and Osteoblast Markers During Bone Development

We examined the expressions of Runx2, osteopontin, and osteocalcin during bone development by immunohistochemistry (Fig. 1). Osteopontin is expressed in relatively immature osteoblasts and osteocalcin is expressed in mature osteoblasts. In the femur, at 1 week of age, Runx2, but not osteopontin and osteocalcin, was expressed in the perichondrial region surrounding proliferating and prehypertrophic chondrocytes. In the perichondrial region surrounding hypertrophic chondrocytes, osteopontin, but not osteoclalcin, was expressed with Runx2. Osteocalcin was first detected in osteoblasts that expressed Runx2 and osteopontin, in the bone collar surrounding primary spongiosa. In the diaphysis at 1 week of age and the metaphysis at 4 weeks of age, the osteoblasts, which expressed Runx2 and osteopontin strongly, weakly expressed osteocalcin; the osteoblasts, which expressed osteocalcin strongly and Runx2 weakly, barely expressed osteopontin. In the diaphysis at 4 weeks of age, Runx2 and osteopontin were barely detectable in osteoblasts, but osteocalcin was strongly expressed in the osteoblasts. Therefore, Runx2 is first detected in preosteoblasts, and the expression increases in immature osteoblasts, reducing during osteoblast maturation; mature osteoblasts do not express a significant amount of Runx2 protein [16] (Fig. 2).

3 Osteoblast Differentiation at an Early Stage

Skeletal component cells, including osteoblasts, chondrocytes, adipocytes, myoblasts, tendon cells, and fibroblasts, are derived from mesenchymal stem cells. Their lineages are determined by different transcription factors. The transcription factors, Runx2 and Sp7/Osterix, and canonical Wnt signaling regulate osteoblast differentiation; Sox family transcription factors (Sox9, Sox5, and Sox6) regulate chondrocyte differentiation; MyoD transcription factors (MyoD, Myf5, and myogenin) regulate myogenic differentiation, and C/EBP family (C/EBPβ, C/EBPδ, and C/EBPα) and PPARγ2 transcription factors regulate adipocyte differentiation [11]. Sp7 belongs to the Sp family of transcription factors, and Sp7-deficient mice also

Fig. 1 (continued) Expression of Runx2, osteopontin, and osteocalcin proteins during bone development. Single labeling with anti-Runx2 antibody (**A**, **D**, **G**, **J**, **M**, **P**), double labeling with anti-Runx2 and anti-osteopontin antibodies (**B**, **E**, **H**, **K**, **N**, **Q**), and double labeling with anti-Runx2 and anti-osteocalcin antibodies (**C**, **F**, **I**, **L**, **O**, **R**) are shown. The single labeling with anti-osteopontin or anti-osteocalcin antibody is not shown here. The labeling with anti-Runx2 antibody is stained *black*, and the labeling with anti-osteopontin or anti-osteocalcin antibody is stained *brown*. The sections were counterstained with methyl *green*. (**A–L**) Serial sections from femurs at 1 week of age. (**A–C**) The perichondrial region surrounding proliferating and prehypertrophic chondrocyte layers is shown. (**D–F**) The perichondrial region surrounding the hypertrophic chondrocyte layer is shown. (**G–L**), Cortical bones at metaphysis (**G–H**) and diaphysis (**J–L**) are shown. (**M–R**) Serial sections from femurs at 4 weeks of age. Cortical bones at metaphysis (**M–O**) and diaphysis (**P–R**) are shown. Bars: 30 μm

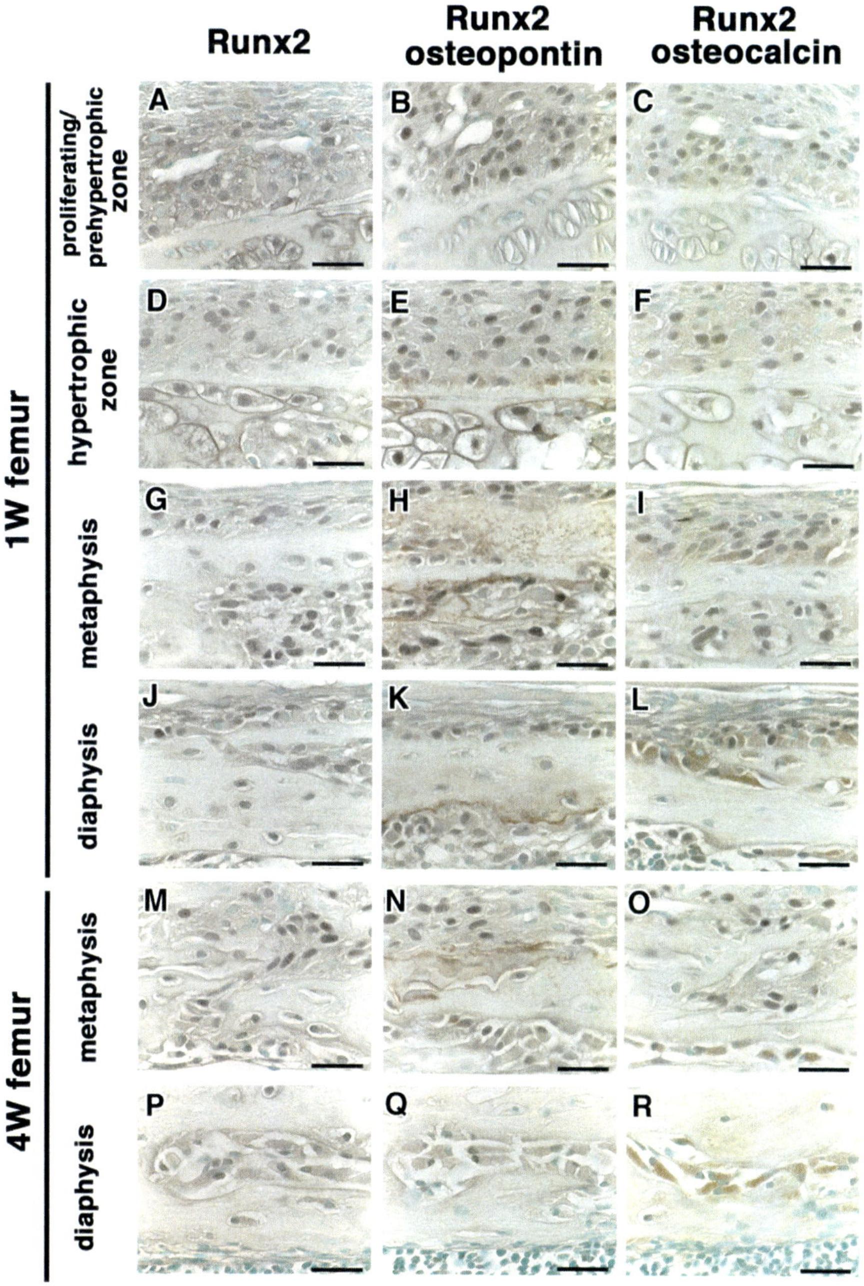

Fig. 1 (continued)

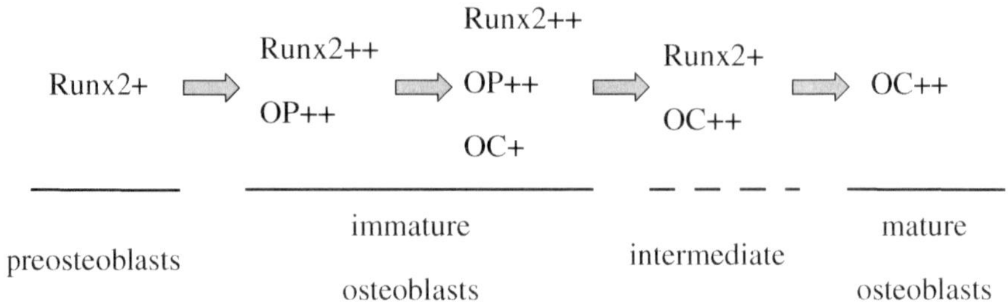

Fig. 2 Runx2 protein expression during osteoblast differentiation. OP: osteopontin, OC: osteocalcin

lack osteoblasts [17]. In Sp7-deficient mice, Runx2 is expressed in the perichondrial cells surrounding hypertrophic chondrocytes, and these cells differentiated into chondrocytes. Recently, several groups reported that Ctnnb1/b-catenin depletion in osteoblast progenitors or in preosteoblasts resulted in a complete block of osteoblast differentiation [1, 5, 6, 19]. In these conditional knockout mice, Runx2 is expressed in the perichondrial cells surrounding hypertrophic chondrocytes, and these cells differentiate into chondrocytes. Sp7 is not detected in perichondrial cells whose Ctnnb1 had been depleted at the osteoblast progenitor stage, whereas Sp7 is detected in perichondrial cells whose Ctnnb1 had been depleted at the preosteoblast stage. This result indicates that canonical Wnt signaling is required for not only Sp7 expression, but also for osteoblast differentiation after the expression of Sp7; therefore, Runx2 is the first transcription factor required for determination of the osteoblast lineage, while Sp7 and canonical Wnt signaling further direct the fate of mesenchymal cells to osteoblasts, blocking their differentiation into chondrocytes (Fig. 3).

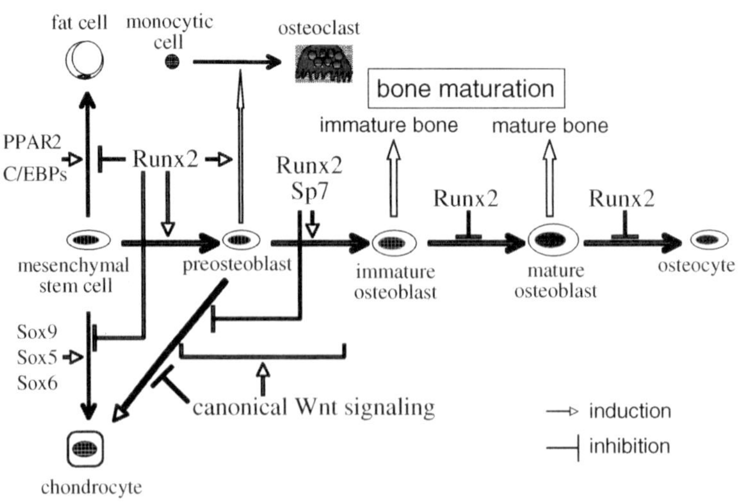

Fig. 3 Regulation of osteoblast differentiation by Runx2, Sp7, and canonical Wnt signaling (from Komori [10])

4 Osteoblast Differentiation at a Late Stage

Runx2 transgenic mice under the control of the 2.3 kb Col1a1 promoter, which directs the transgene expression to immature and mature osteoblasts, show severe osteopenia with multiple fractures. Osteoblasts, which express osteopontin, are accumulated but osteoblasts, which express osteocalcin, are decreased, indicating that osteoblast maturation is inhibited in transgenic mice. The cortical bone of adult transgenic mice is composed of woven bone and easily resorbed by osteoclasts [15]. Furthermore, addition of the Cbfb transgene in Runx2 transgenic mice worsens osteopenia [8]; therefore, Runx2 inhibits osteoblast maturation. In dominant-negative (dn) Runx2 transgenic mice under the same 2.3 kb Col1a1 promoter, the trabecular bone is increased and the collagen fibrils are more densely packed and more highly organized, when compared with wild-type mice, and mineralization is greater than that of wild-type mice, leading to less bone resorption (Fig. 4). Additionally, severe reduction in the number of osteocytes in Runx2

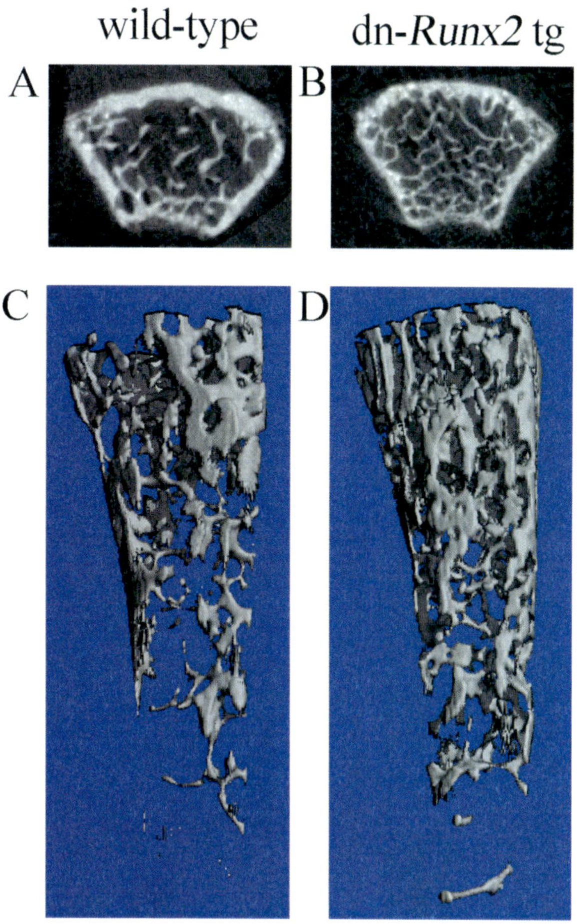

Fig. 4 Micro-CT analysis. Femurs from wild-type (**A**, **C**) and dn-Runx2 transgenic (tg) (**B**, **D**) mice at 13 weeks of age were analyzed by micro-CT. Two-dimensional axial image of distal femoral metaphysis (**A**, **B**) and three-dimensional trabecular bone architecture of distal femoral metaphysis (**C**, **D**) are shown

transgenic mice is restored in Runx2/dn-Runx2 double transgenic mice [16]. These findings indicate that Runx2 inhibits osteoblast maturation and induces immature bone formation, and that Runx2 has to be suppressed for mature bone formation. In accordance with these findings, the level of Runx2 protein was reduced in mature osteoblasts (Fig. 1) [16].

5 Regulation of Bone Matrix Protein Gene Expression by Runx2

The DNA-binding sites of Runx2 in major bone matrix protein genes, including Col1a1; Col1a2; Spp1; Ibsp/BSP; Bglap2; Fn1/fibronectin; Mmp13, and Tnfrsf11b/Opg, have been identified, and Runx2 induced the expression of these genes or activated their promoters in vitro [3, 4, 7, 9, 14, 20]. Furthermore, the major bone matrix protein gene expressions were markedly reduced in dn-Runx2 tranegenic mice under the control of osteocalcin promoter [2]. In the osteoblasts of Col1a1 promoter Runx2 transgenic mice, however, the expression of Col1a1, alkaline phosphatase (Akp2), Bglap2, and Mmp13 – all of which normally increase during osteoblast maturation – are reduced [15]. Furthermore, dn-Runx2 fails to reduce Col1a1, Spp1, and Bglap2 in mature osteoblasts, in vitro and in vivo [16]; therefore, Runx2 regulates the expression of major bone matrix genes during the early stage of osteoblast differentiation, but Runx2 is not essential to maintain these gene expressions in mature osteoblasts. This is also compatible with the expression pattern of Runx2 in osteoblasts, in which Runx2 is downregulated in mature osteoblasts, during bone development (Fig. 1) [16].

References

1. Day, T.F., Guo, X., Garrett-Beal, L., & Yang, Y. (2005). Wnt/β-catenin signaling in mesenchymal progenitors controls osteoblast and chondrocyte differentiation during vertebrate skeletogenesis. *Dev Cell*, 8, 739–750
2. Ducy, P., Starbuck, M., Priemel, M., Shen, J., Pinero, G., Geoffroy, V., Amling, M., & Karsenty, G. (1999). A Cbfa1-dependent genetic pathway controls bone formation beyond embryonic development. *Genes Dev*, 13, 1025–1036
3. Ducy, P., Zhang, R., Geoffroy, V., Ridall, A.L., & Karsenty, G. (1997). Osf2/Cbfa1: a transcriptional activator of osteoblast differentiation. *Cell*, 89, 747–754
4. Harada, H., Tagashira, S., Fujiwara, M., Ogawa, S., Katsumata, T., Yamaguchi, A., Komori, T., & Nakatsuka, M. (1999). Cbfa1 isoforms exert functional differences in osteoblast differentiation. *J Biol Chem*, 274, 6972–6978
5. Hill, T.P., Später, D., Taketo, M.M., Birchmeier, W., & Hartmann, C. (2005). Canonical Wnt/?-catenin signaling prevents osteoblasts from differentiating into chondrocytes. *Dev Cell*, 8, 727–738
6. Hu, H., Hilton, M.J., Tu, X., Yu, K., Ornitz, D.M., & Long, F. (2005). Sequential roles of Hedgehog and Wnt signaling in osteoblast development. *Development*, 132, 49–60
7. Jiménez, M.J., Balbín, M., López, J.M., Alvarez, J., Komori, T., & López-Otín, C. (1999). Collagenase 3 is a target of Cbfa1, a transcription factor of the runt gene family involved in bone formation. *Mol Cell Biol*, 19, 4431–4442

8. Kanatani, N., Fujita, T., Fukuyama, R., Liu, W., Yoshida, C.A., Moriishi, T., Yamana, K., Miyazaki, T., Toyosawa, S., & Komori, T. (2006). Cbfb regulates Runx2 function isoform-dependently in postnatal bone development. *Dev Biol*, 296, 48–61
9. Kern, B., Shen, J., Starbuck, M., & Karsenty, G. (2001). Cbfa1 contributes to the osteoblast-specific expression of type I collagen genes. *J Biol Chem*, 276, 7101–7107
10. Komori, T. (2008). Regulation of bone development and maintenance by Runx2. *Front Biosci*, 13, 898–903
11. Komori, T. (2006). Regulation of osteoblast differentiation by transcription factors. *J Cell Biochem*, 99, 1233–1239
12. Komori, T. (2005). Regulation of skeletal development by the Runx family of transcription factors. *J Cell Biochem*, 95, 445–453
13. Komori, T., Yagi, H., Nomura, S., Yamaguchi, A., Sasaki, K., Deguchi, K., Shimizu, Y, Bronson, R.T., Gao, Y.H., Inada, M., Sato, M., Okamoto, R., Kitamura, Y., Yoshiki, S., & Kishimoto, T. (1997). Targeted disruption of Cbfa1 results in a complete lack of bone formation owing to maturational arrest of osteoblasts. *Cell*, 89, 755–764
14. Lee, K.S., Kim, H.J., Li, Q.L., Chi, X.Z., Ueta, C., Komori, T., Wozney, J.M., Kim, E.G., Choi, J.Y., Ryoo, H.M., & Bae, S.C. (2000). Runx2 is a common target of transforming growth factor b1 and bone morphogenetic protein 2, and cooperation between Runx2 and Smad5 induces osteoblast-specific gene expression in the pluripotent mesenchymal precursor cell line C2C12. *Mol Cell Biol*, 20, 8783–8792
15. Liu, W., Toyosawa, S., Furuichi, T., Kanatani, N., Yoshida, C., Liu, Y., Himeno, M., Narai, S., Yamaguchi, A., & Komori, T. (2001). Overexpression of Cbfa1 in osteoblasts inhibits osteoblast maturation and causes osteopenia with multiple fractures. *J Cell Biol*, 155, 157–166
16. Maruyama, Z., Yoshida, C.A., Furuichi, T., Amizuka, N., Ito, M., Fukuyama, R., Miyazaki, T., Kitaura, H., Nakamura, K., Fujita, T., Kanatani, N., Moriishi, T., Yamana, K., Liu, W., Kawaguchi, H., Nakamura, K., & Komori, T. (2007). Runx2 determines bone maturity and turnover rate in postnatal bone development and is involved in bone loss in estrogen deficiency. *Dev Dyn*, 236, 1876–1890
17. Nakashima, K., Zhou, X., Kunkel, G., Zhang, Z., Deng, J.M., Behringer, R.R., & de Crombrugghe, B. (2002). The novel zinc finger-containing transcription factor Osterix is required for osteoblast differentiation and bone formation. *Cell*, 108, 17–29
18. Otto, F., Thornell, A.P., Crompton, T., Denzel, A., Gilmour, K.C., Rosewell, I.R., Stamp, G.W., Beddington, R.S., Mundlos, S., Olsen, B.R., Selby, P.B., & Owen, M.J. (1997). Cbfa1, a candidate gene for cleidocranial dysplasia syndrome, is essential for osteoblast differentiation and bone development. *Cell*, 89, 765–771
19. Rodda, S.J., & McMahon, A.P. (2006). Distinct roles for Hedgehog and canonical Wnt signaling in specification, differentiation and maintenance of osteoblast progenitors. *Development*, 133, 3231–3244
20. Thirunavukkarasu, K., Halladay, D.L., Miles, R.R., Yang, X., Galvin, R.J., Chandrasekhar, S., Martin, T.J., & Onyia, J.E. (2000). The osteoblast-specific transcription factor Cbfa1 contributes to the expression of osteoprotegerin, a potent inhibitor of osteoclast differentiation and function. *J Biol Chem*, 275, 25163–25172
21. Yoshida, C.A., Furuichi, T., Fujita, T., Fukuyama, R., Kanatani, N., Kobayashi, S., Satake, M., Takada, K., & Komori, T. (2002). Core-binding factor b interacts with Runx2 and is required for skeletal development. *Nat Genet*, 32, 633–638

Communication Between EphrinB2 and EphB4 Within the Osteoblast Lineage

T.J. Martin, E.H. Allan, P.W.M. Ho, J.H. Gooi, J.M.W. Quinn, M.T. Gillespie, V. Krasnoperov, and N.A. Sims

Abstract Members of the ephrin and Eph family are local mediators of cell function through largely contact-dependent processes in development and in maturity. Production of ephrinB2 mRNA and protein are increased by PTH and PTHrP in osteoblasts. Both a synthetic peptide antagonist of ephrinB2/EphB4 receptor interaction and recombinant soluble extracellular domain of EphB4 (sEphB4), which is an antagonist of both forward and reverse EphB4 signaling, were able to inhibit mineralization and the expression of several osteoblast genes involved late in osteoblast differentiation. The findings are consistent with ephrinB2/EphB4 signaling within the osteoblast lineage having a paracrine role in osteoblast differentiation, in addition to the proposed role of osteoclast-derived ephrinB2 in coupling of bone formation to resorption. This local regulation might contribute to control of osteoblast differentiation and bone formation at remodeling sites, and perhaps also in modeling.

Keywords Osteoblast · Ephrins · PTHrP · Bone remodeling

1 Introduction – Bone Remodeling

The maintenance of adequate trabecular and cortical bone requires that bone formation and resorption should be balanced, such that a high or low level of resorption is usually associated with a similar change in the level of bone formation. Bone resorption and formation take place asynchronously throughout the skeleton in both trabecular and cortical bone, at many sites known as bone metabolic units (BMUs). The theory that resorption is followed by an equal amount of formation in the BMU came to be known as "coupling." There are a number of situations in which this equal balance does not hold. During aging there is a negative balance at individual

---

T.J. Martin (✉)
Department of Medicine, St Vincent's Institute and University of Melbourne, 9 Princes Street, Melbourne, Fitzroy, 3065, Australia
e-mail: jmartin@svi.edu.au

Y. Choi (ed.), *Osteoimmunology*, Advances in Experimental Medicine and Biology 658, DOI 10.1007/978-1-4419-1050-9_6, © Springer Science+Business Media, LLC 2010

BMUs [14] with gradual diminution in the amount of bone, whereas during growth it is proposed that there is a positive balance, with the amount of bone replaced at individual BMUs exceeding that which was lost [23]. In the states of post-menopausal osteoporosis or of ovariectomy in animal models, coupling is perturbed to the extent that the bone resorbed within each BMU is inadequately replaced by formation; the net result being bone loss. The tightly regulated processes of bone formation and resorption are essential in bone remodeling, for the achievement and maintenance of skeletal strength and form. Circulating hormones are important controlling factors, but the key influences are locally generated cytokines, which influence bone cell function and communication in complex ways, and often are themselves regulated in turn by the hormones.

Bone remodeling is essential in order to repair microdamage to bone and to respond to pressure changes [22]. The bone-remodeling sequence begins with signals generated from cells in the osteoblast lineage, osteocytes and bone lining cells, aimed at generating active osteoclasts to resorb old or damaged bone [2, 20]. This is followed by the reversal phase on completion of bone resorption, with the death and departure of multinucleate osteoclasts, when the resorbed surface is cleared by mononuclear cells, probably both macrophage and mesenchymal in origin [6]. In the bone-formation phase, the resorbed bone is replaced by the actions of osteoprogenitor cells, which differentiate into osteoblasts. Uncalcified matrix (osteoid) is deposited for subsequent mineralization. The lacuna is gradually filled with new bone. These processes clearly require controlled production of molecules by the participating cells and their communication with neighbours.

2 Communication from the Osteoblast to the Hemopoietic Lineage

The crucial local factors that control osteoclast formation were discovered in the late 1990s. Osteoblasts express a membrane protein, Receptor Activator of NF-κB ligand (RANKL) regulated by osteotropic hormones, including parathyroid hormone and calcitriol, as well as cytokines such as interleukin-6 [27]. RANKL plays an essential role in osteoclast differentiation, activation, and survival. Proximity between osteoblastic lineage and hemopoietic cells is required for RANKL to bind to its respective receptor (RANK), which is expressed by monocyte/macrophage lineage cells, thereby stimulating some of these to form osteoclasts. The binding of RANKL to its receptor in mononuclear hematopoietic precursors initiates the processes that ultimately lead to the formation of multinucleate osteoclasts. Osteoprotegerin (OPG) acts as a decoy receptor for RANKL to suppress osteoclast formation. Studies in genetically altered mice have clearly established the essential physiological roles of RANKL and OPG in controlling osteoclast formation and activity, and greatly enhanced our understanding of this stage of the bone-remodeling process – the early recruitment of osteoclast precursors, and the important role played in this by cells of the osteoblast lineage.

3 Communication from Osteoclasts to the Osteoblasts in Order to Contribute to Bone Formation

Much less is known of communication in the reverse direction, even though it has long been recognized that rates of bone resorption are generally matched by those of bone formation. A local "coupling factor" linking bone resorption to subsequent formation was proposed as the key regulator of the remodeling process [17]. The concept developed that coupling might be achieved by the activities of one or more growth factors released from bone matrix during resorption, with most credence given to IGF I and II, and TGFβ [3, 21]. This model of coupling in the BMU by growth factor release from the matrix raises a number of questions that relate to the time course and the distance between the resorption and formation processes and whether activation can be controlled with sufficient precision: (i) which cells produce the growth factors and under what circumstances; (ii) do they stimulate bone formation in vivo; (iii) which can be released from the matrix in active form and in a spatially and temporally controlled manner; (iv) is there evidence for an increase in the abundance of these substances at sites of bone remodeling; and, (v) are there regulated mechanisms by which they are activated?

On the other hand, it is possible that coupling of bone formation might be achieved through activities generated from active osteoclasts. Evidence consistent with this came from Nakamura et al. [19], who used osteoprotegerin $(OPG)^{-/-}$ mice, which are severely osteoporotic because of excessive osteoclast formation, to show that these mice have greatly increased bone formation resulting from a local active factor. They suggested that this factor is more likely derived from cells than released from matrix [27]. Other evidence came from studies in mice, in which each of the two gp130-dependent signaling pathways was specifically attenuated. Inactivation of the SHP2/ras/MAPK signaling pathway ($gp130^{Y757F/Y757F}$ mice) yielded mice with greater osteoclast numbers and bone resorption, as well as greater bone formation than wild type mice. This increased bone remodeling resulted in less bone because the increase in resorption was relatively greater than that in formation. When $gp130^{Y757F/Y757F}$ mice were crossed with IL-6 null mice they had similarly high osteoclast numbers and increased bone resorption; however, these mice showed no corresponding increase in bone formation and thus had extremely low bone mass. This indicated that stimulation of bone formation coupled to the high level of bone resorption in $gp130^{Y757F/Y757F}$ mice is of cellular rather than resorbed matrix origin, and is an IL-6-dependent process, though it does not necessarily show that it is mediated by IL-6 itself [26].

The first specific mechanism proposed as an osteoclast – derived mediator of coupling was ephrinB2, with the finding that osteoclast-derived ephrinB2 acts through a contact-dependent mechanism on EphB4, its receptor in the osteoblast, to promote osteoblast differentiation and bone formation [30]. This was of interest because ephrin/Eph family members have been recognized for some time as local mediators of cell function through largely contact-dependent processes in development and in maturity [5, 10, 24]. They mediate cell attraction and adhesion, but often also provide signals that separate the cells, and have demonstrated roles in tissue remodeling, including cell migration, vascular development, axon guidance and synapse plasticity. Based on the ephrin ligand structure, the ephrins are in two classes, with the ephrin A class glycophosphatidylinositol (GPI)-tethered to the membrane and the B class consisting of type II membrane proteins [10, 24]. Although first considered to bind with class-specificity, i.e. ephrinA to EphA and ephrinB to EphB, there are exceptions, with EphA4 being activated through ephrinB1 [11] and ephrinA5 binding to and signaling through EphB2 [9, 18]. A particular feature of ephrin/Eph biology is their capacity for bi-directional signaling, in that when an ephrin acts upon its Eph receptor tyrosine kinase, the latter can signal in the reverse direction, acting through the ligand by promoting rapid phosphorylation on highly conserved tyrosine residues within the cytoplasmic tail [15].

Transgenic mice constructed to overexpress EphB4 in the osteoblast lineage showed increased bone formation parameters, and treatment of osteoblastic cells in vitro with ephrinB2-Fc fusion protein resulted in increased expression of genes associated with osteoblast differentiation [30]. By transfecting osteoblasts with ephrinB2 mutant forms it was concluded also that the action of ephrinB2 did not require its participation in intracellular signaling in the osteoblasts, but rather that the ephrinB2 extracellular domain stimulated the EphB4 receptors. Of great interest also in this work was the evidence that through reverse signaling, osteoblast-derived EphB4 could act upon ephrinB2 in osteoclasts to suppress osteoclast differentiation by inhibiting the cFos/NFATc1 cascade that is essential for osteoclast differentiation.

The concept of ephrin-Eph involvement in the coupling process is an intriguing one. The fact that these interactions seem to require cell contact between the osteoclast and a differentiating osteoblast makes it likely that it could be just one of many contributory factors to coupling. The findings to be discussed below extend the involvement of ephrinB2 to indicate that it might also be involved in remodeling in ways that determine the amount of bone formed in the BMU.

4 How Do Cells of the Osteoblast Lineage Know How Much Bone to Make in a BMU?

Among the many unanswered questions concerning bone remodeling is why osteoclasts stop resorbing after excavating a certain amount of bone, and either die or move on. So, it would also be intriguing to understand how the osteoblast lineage cells, differentiating to form bone within the BMU, do so virtually precisely to replace the amount of bone that has been lost. An interesting insight into this comes from the work of Boyde and colleagues [7] who showed in vitro that, if they provided rat calvarial cells to bone slices with crevices and grooves excavated on them, the production of bone was limited to the space available. Their findings suggested that the topography of the bone affected the timing, siting and extent of new bone formation, and that in vivo this would take place in the resorbed spaces vacated by osteoclasts. Both the proposed growth factor involvement and the work of Gray et al. imply that once the formation process is established, the participating cells themselves are able to sense spatial limits, and most likely do so by chemical communication within that population of cells. Growth factors and cytokines produced by those cells are candidate mediators, as are gap junction intercellular communicators. Relevant to the latter, deletion of connexin43 has been shown to result in impaired response of bone formation to loading and to treatment with PTH [4]. Regulation from outside the BMU population could be provided by osteocyte-derived sclerostin, which could communicate with those cells to limit bone formation [25, 28].

Another possible regulatory influence arises from our finding that production of ephrinB2 by osteoblasts is substantially enhanced by PTH and PTHrP [1]. In using gene profiling to study the actions of PTH(1-34) and PTHrP(1-141) on differentiating osteoblasts, we found that among the ephrin and Eph molecules represented on the array, ephrinB2 mRNA and protein production was substantially enhanced by treatment with either PTH or PTHrP (Fig. 1). This effect was confirmed in UMR106 rat osteogenic sarcoma cells (Fig. 1) and in mouse calvarial osteoblasts [1]. The response was a rapid one, evident within 1 h, and the increased ephrinB2 protein production was maintained for greater than 24 h (Fig. 1). Furthermore treatment of either 3-week old rats or ovariectomized 6-month old rats with a single subcutaneous injection of PTH resulted in a tenfold increase in ephrinB2 mRNA in metaphyseal bone [1]. Localization of ephrinB2 by immunohistology in young rat bone showed it in clusters of osteoblasts, predominantly in association with mature rather than woven bone [1], consistent with ephrinB2 involvement in remodeling.

Since clustered ephrinB2 treatment of mouse calvarial osteoblasts has been shown to enhance expression of genes associated with osteoblast differentiation, and mice overexpressing EphB4 have increased bone formation parameters, we used inhibitors of ephrinB2-EphB4 receptor interaction in in vitro experiments to determine whether, within a population of osteoblastic cells, they influenced mineralization of differentiating osteoblasts or expression of differentiation-related genes. Two classes of receptor antagonist were used. The first was the peptide, TNYLFSPNGPIARAW, discovered through phage display and shown to be a specific antagonist of ephrinB2 interaction with EphB4 [13]. The second was the recombinant extracellular domain of EphB4 (sEphB4), shown to inhibit both forward and reverse signaling between ephrinB2 and EphB4 [12]. Each of these receptor antagonists was able to inhibit mineralization of the mouse stromal cells, Kusa 4b10, in a dose-dependent manner (Fig. 2A). Furthermore, in Kusa 4b10 cells at a late stage of differentiation, sEphB4 inhibited the expression of mRNA for a number of genes associated with osteoblast differentiation (Fig. 2B), similar to our previous findings with the peptide receptor antagonist, TNYL [1]. Additionally, in mouse calvarial osteoblasts differentiated over 7 days in conditions that result in several hundredfold increase in expression of osteocyte markers, both sEphB4 and TNYL over 24 h decreased expression of mRNA for osteocalcin, sclerostin, DMP-1 and MEPE (Fig. 2C). These findings are consistent with a role for ephrinB2 interaction with EphB4 within the osteoblast lineage, quite distinct from the proposed

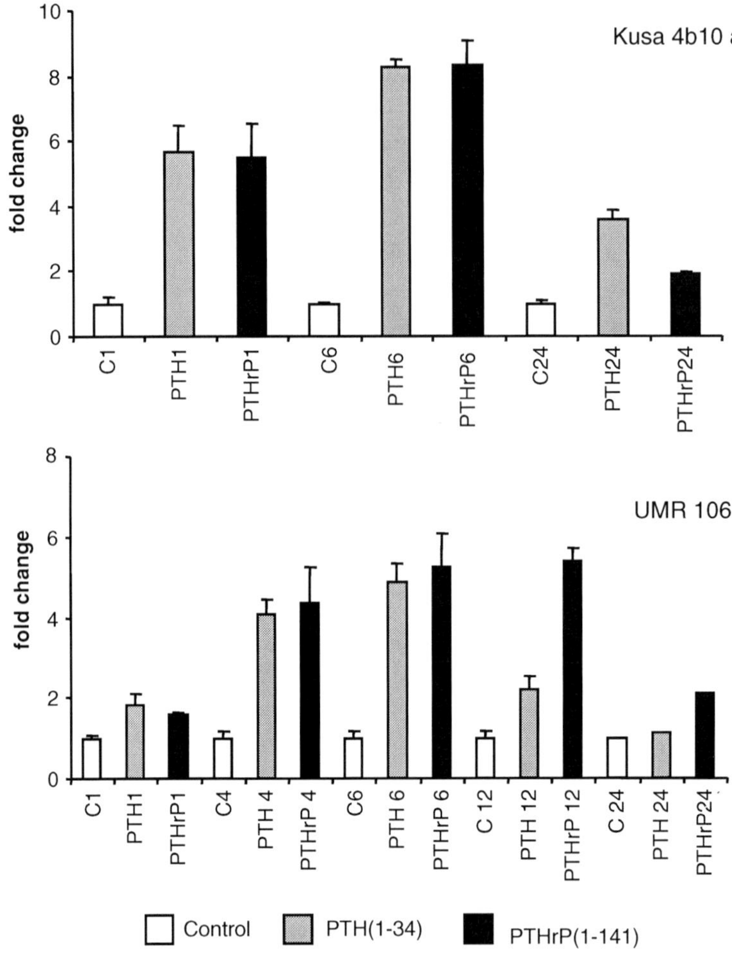

Fig. 1 EphrinB2 mRNA in response to treatment with PTH(1-34) or PTHrP(1-141) for up to 24 h in differentiated Kusa 4b10 cells (*top panel*) or UMR106 rat osteogenic sarcoma cells (*bottom panel*)

role of osteoclast-derived ephrinB2. The communication process is illustrated in Fig. 3, depicting ephrin-Eph interaction playing a role in communication within the osteoblast population in the bone formation phase of bone remodeling.

With locally produced PTHrP in bone as the likely ligand for the PTH receptor in the osteoblast lineage [16], we propose that production of ephrinB2 in osteoblasts is enhanced by activation of the PTH receptor, probably through the paracrine action of PTHrP. A contact-dependent process operating within a group of maturing osteoblasts in the BMU might present ephrinB2 with easier access to its receptor than relying on osteoblast-osteoclast juxtaposition. Nevertheless, ephrinB2/EphB4

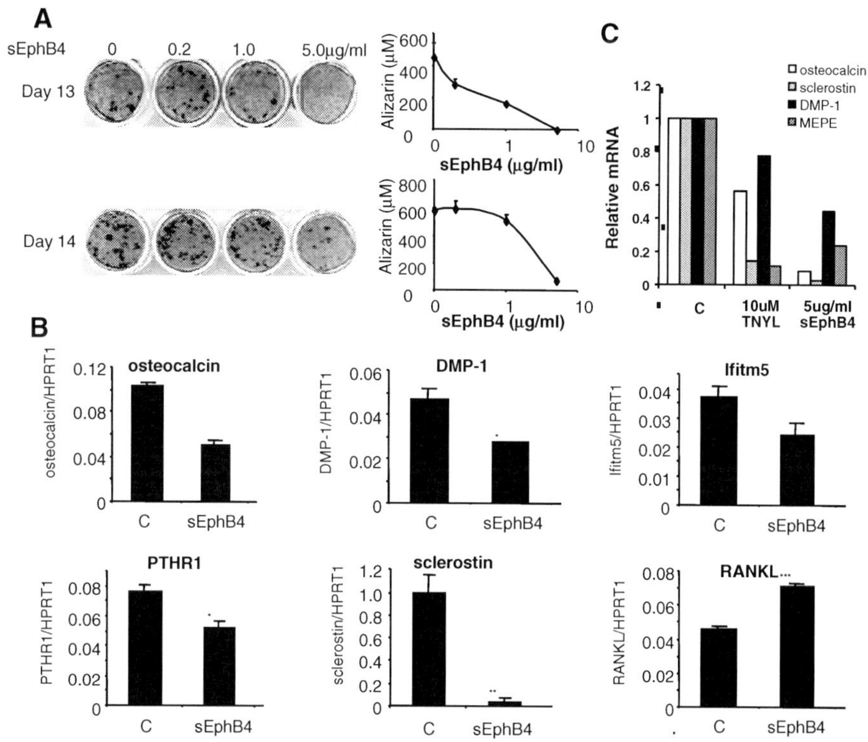

Fig. 2 (**A**) Kusa 4b10 cells under mineralizing conditions treated with increasing concentrations of sEphB4 and mineralization measured at 13 and 14 days; (**B**) Kusa 4b10 cells, mRNA for several genes after 24-h treatment with sEphB4 (5 μg/ml) or control; (**C**) mouse calvarial osteoblasts after differentiation for 7 days, and 24-h treatment with either peptide antagonist (TNYL) or sEphB4

Fig. 3 Osteoblasts depicted filling a remodeling space, with ephrinB2 – EphB4 interactions shown, as well as intercellular communication by gap junctions and signals from matrix

forward and reverse signaling is an attractive means of regulating the volumes of bone resorbed and formed.

These considerations of ephrin-Eph interactions in bone are thus far limited to observations made with bone cells. The abundance of important ephrin signaling processes in the vasculature [29] make it imperative to determine whether that source also applies in the process of bone remodeling. In proposing the model of intercellular communication within the osteoblast lineage that we report here, it will be essential to evaluate the relative importance of forward and reverse signaling. For example, ephrinB2 forward signaling and EphB4 reverse signaling were found to affect cell adhesion and migration differentially between arterial and venous endothelial cells [8]. Although our data in osteoblasts taken together taken with that of Zhao et al., in osteoclast-osteoblast communication might favor actions through ephrinB2 forward signaling, the possibility of significant effects through EphB4 reverse signaling has not been excluded. Resolution of this and related questions will require further in vivo and ex vivo studies in genetically manipulated mice.

5 Summary

During bone remodeling there are contributions from many important pathways of intercellular communication among the osteoblast lineage, osteoclasts and cells of the immune system. Although coupling of formation to resorption in bone remodeling is often ascribed to a hypothetical "coupling factor," the overall mechanism seems too complex to be explained by any single factor, and it is likely that there are several contributors. One of the important stages in remodeling is the filling of the excavated space in the BMU, where differentiation of osteoblast precursors must be controlled and the amount of bone replaced must be limited. The finding of ephrinB2 regulation within this lineage by a local modulator of bone remodeling, PTHrP, introduces this pathway as another that might influence the process of bone renewal.

References

1. Allan, E.H., Hausler, K.D., Wei, T., Gooi, J.H., Quinn, J.M., Crimeen-Irwin, B., Pompolo, S., Sims, N.A., Gillespie, M.T, Onyia, J.E., & Martin, T.J. (2008). EphrinB2 Regulation by Parathyroid Hormone (PTH) and PTHrP Revealed by Molecular Profiling in Differentiating Osteoblasts. *J Bone Miner Res*, 23(8), 1170–1181.
2. Bonewald, L.F. (2007). Osteocyte messages from a bony tomb. *Cell Metab*, 5(6), 410–411.
3. Centrella, M., McCarthy, T.L., & Canalis, E. (1991). Transforming growth factor-beta and remodeling of bone. *J Bone Joint Surg Am*, 73(9), 1418–1428.
4. Civitelli, R. (2008). Cell-cell communication in the osteoblast/osteocyte lineage. *Arch Biochem Biophys*, 473(2), 188–192.
5. Compagni, A., Logan, M., Klein, R., & Adams, R.H. (2003). Control of skeletal patterning by ephrinB1-EphB interactions. *Dev Cell*, 5(2), 217–230.

6. Everts, V., Delaisse, J.M., Korper, W., Jansen, D.C., Tigchelaar-Gutter, W., Saftig, P., & Beertsen, W. (2002). The bone lining cell: its role in cleaning Howship's lacunae and initiating bone formation. *J Bone Miner Res*, 17(1), 77–90.
7. Gray, C., Boyde, A., & Jones, S.J. (1996). Topographically induced bone formation in vitro: implications for bone implants and bone grafts. *Bone*, 18(2), 115–123.
8. Hamada, K., Oike, Y., Ito, Y., Maekawa, H., Miyata, K., Shimomura, T., & Suda, T. (2003). Distinct roles of ephrin-B2 forward and EphB4 reverse signaling in endothelial cells. *Arterioscler Thromb Vasc Biol*, 23(2), 190–197.
9. Himanen, J.P., Chumley, M.J., Lackmann, M., Li, C., Barton, W.A., Jeffrey, P.D., Vearing, C., Geleick, D., Feldheim, D.A., Boyd, A.W., Henkemeyer, M., & Nikolov, D.B. (2004). Repelling class discrimination: ephrin-A5 binds to and activates EphB2 receptor signaling. *Nat Neurosci*, **7**(5):501–509.
10. Himanen, J.P., & Nikolov, D.B. (2003). Eph receptors and ephrins. *Int J Biochem Cell Biol*, 35(2), 130–134.
11. Holmberg, J., Armulik, A., Senti, K.A., Edoff, K., Spalding, K., Momma, S., Cassidy, R., Flanagan, J.G., & Frisen, J. (2005). Ephrin-A2 reverse signaling negatively regulates neural progenitor proliferation and neurogenesis. *Genes Dev*, 19(4), 462–471.
12. Kertesz, N., Krasnoperov, V., Reddy, R., Leshanski, L., Kumar, S.R., Zozulya, S., & Gill, P.S. (2006). The soluble extracellular domain of EphB4 (sEphB4) antagonizes EphB4-EphrinB2 interaction, modulates angiogenesis, and inhibits tumor growth. *Blood*, 107(6), 2330–2338.
13. Koolpe, M., Burgess, R., Dail, M., & Pasquale, E.B. (2005). EphB receptor-binding peptides identified by phage display enable design of an antagonist with ephrin-like affinity. *J Biol Chem*, 280(17), 17301–17311.
14. Lips, P., Courpron, P., & Meunier, P.J. (1978). Mean wall thickness of trabecular bone packets in the human iliac crest: changes with age. *Calcif Tissue Res*, 26(1), 13–17.
15. Lu Q, Sun E.E., Klein R.S., Flanagan J.G. (2001). Ephrin-B reverse signaling is mediated by a novel PDZ-RGS protein and selectively inhibits G protein-coupled chemoattraction. *Cell*, 105(1), 69–79.
16. Miao, D., He, B., Jiang, Y., Kobayashi, T., Soroceanu, M.A., Zhao, J., Su, H., Tong, X., Amizuka, N., Gupta, A., Genant, H.K., Kronenberg, H.M., Goltzman, D., & Karaplis, A.C. (2005). Osteoblast-derived PTHrP is a potent endogenous bone anabolic agent that modifies the therapeutic efficacy of administered PTH 1-34. *J Clin Invest*, 115(9), 2402–2411.
17. Mohan, S., & Baylink, D.J. (1991). Bone growth factors. *Clin Orthop Relat Res*, 263, 30–48.
18. Murai, K.K., & Pasquale, E.B. (2003). 'Eph'ective signaling: forward, reverse and crosstalk. *J Cell Sci*, 116(Pt 14), 2823–2832.
19. Nakamura, M., Udagawa, N., Matsuura, S., Mogi, M., Nakamura, H., Horiuchi, H., Saito, N., Hiraoka, B.Y., Kobayashi, Y., Takaoka, K., Ozawa, H., Miyazawa, H., & Takahashi, N. (2003). Osteoprotegerin regulates bone formation through a coupling mechanism with bone resorption. *Endocrinology*, 144(12), 5441–5449.
20. Noble, B.S., Peet, N., Stevens, H.Y., Brabbs, A., Mosley, J.R., Reilly, G.C., Reeve, J., Skerry, T.M., & Lanyon, L.E. (2003). Mechanical loading: biphasic osteocyte survival and targeting of osteoclasts for bone destruction in rat cortical bone. *Am J Physiol Cell Physiol*, 284(4), C934–C943.
21. Oreffo, R.O., Mundy, G.R., Seyedin, S.M., & Bonewald, L.F. (1989). Activation of the bone-derived latent TGF beta complex by isolated osteoclasts. *Biochem Biophys Res Commun*, 158(3), 817–823.
22. Parfitt, A.M. (1996). Skeletal heterogeneity and the purposes of bone remodeling: implications for the understanding of osteoporosis. In: Marcus R, Feldman D, Kelsey J (eds.) Osteoporosis. Academic Press, San Diego, CA, pp. 315–339.
23. Parfitt, A.M. (2002). Targeted and nontargeted bone remodeling: relationship to basic multicellular unit origination and progression. *Bone*, 30(1), 5–7.
24. Pasquale, E.B. (2005). Eph receptor signaling [stet] casts a wide net on cell behaviour. *Nat Rev Mol Cell Biol*, 66, 462–475.

25. Robling, A.G., Bellido, T., & Turner, C.H. (2006). Mechanical stimulation in vivo reduces osteocyte expression of sclerostin. *J Musculoskelet Neuronal Interact*, 6(4), 354.
26. Sims, N.A., Jenkins, B.J., Quinn, J.M., Nakamura, A., Glatt, M., Gillespie, M.T., Ernst, M., & Martin, T.J. (2004). Glycoprotein 130 regulates bone turnover and bone size by distinct downstream signaling pathways. *J Clin Invest*, 113(3), 379–389.
27. Suda, T., Takahashi, N., Udagawa, N., Jimi, E., Gillespie, M.T., & Martin, T.J. (1999). Modulation of osteoclast differentiation and function by the new members of the tumor necrosis factor receptor and ligand families. *Endocr Rev*, 20(3), 345–357.
28. van Bezooijen, R.L., ten Dijke, P., Papapoulos, S.E., & Lowik, C.W. (2005). SOST/sclerostin, an osteocyte-derived negative regulator of bone formation. *Cytokine Growth Factor Rev*, 16(3), 319–327.
29. Yancopoulos, G.D., Davis, S., Gale, N.W., Rudge, J.S., Wiegand, S.J., & Holash, J. (2000). Vascular-specific growth factors and blood vessel formation. *Nature*, 407(6801), 242–248.
30. Zhao, C., Irie, N., Takada, Y., Shimoda, K., Miyamoto, T., Nishiwaki, T., Suda, T., & Matsuo, K. (2006). Bidirectional ephrinB2-EphB4 signaling controls bone homeostasis. *Cell Metab*, 4(2), 111–121.

The Unexpected Link Between Osteoclasts and the Immune System

Hiroshi Takayanagi

Abstract Osteoimmunology is an interdisciplinary research field focused on the molecular understanding of the interplay of the skeletal and immune systems. In particular, the interaction between immune cells and osteoclasts is a current major topic of critical interest in this field. The macrophage-osteoclast interaction has long been studied, and the T-cellosteoclast interaction also attracted much attention in the study of arthritis. However, recent reports have revealed a hitherto unknown link between osteoclasts and other immune cells, including B cells and dendritic cells, suggesting a larger number of molecules are in fact shared by osteoclasts and immune cells. These findings will lead to a better understanding of the pathogenesis of diseases affecting both systems and may/will provide a molecular basis for novel therapeutic strategies.

Keywords Osteoclast · T cells · RANKL · NFATc1 · Cathepsin K

1 Introduction

A close relationship between the immune and skeletal systems has been noted for a long time. For example, the interaction between monocytes/macrophages and osteoclasts was appreciated in the pioneering works performed in the early 1970s [4, 10]: soluble factors secreted from antigen-stimulated peripheral blood mononuclear cells came to be known as osteoclast-activating factors, one of which was revealed to be interleukin-1 (IL-1)[3]. It is now well known that numerous macrophage-derived cytokines influence osteoclast differentiation and/or function. In addition,

H. Takayanagi (✉)
Department of Cell Signaling, Graduate School, Tokyo Medical and Dental University, Yushima 1-5-45, Bunkyo-ku, Tokyo 113-8549, Japan
e-mail: taka.csi@tmd.ac.jp

Y. Choi (ed.), *Osteoimmunology*, Advances in Experimental Medicine and Biology 658, DOI 10.1007/978-1-4419-1050-9_7,

osteoclasts are derived from the monocyte/macrophage lineage and macrophage-colony stimulating factor (M-CSF), which is essential for the development and survival of the lineage, and indispensable for osteoclast differentiation. Therefore, the importance of monocytes/macrophages in osteoclast biology is well established.

The interaction between T cells and osteoclasts is seen in the abnormal activation of the immune system in certain autoimmune diseases, such as rheumatoid arthritis, an aberration which leads to bone destruction. The major cytokine that regulates osteoclast differentiation is receptor activator of NF-κB lingad (RANKL). RANKL is not only expressed by the osteoblasts that support osteoclastogenesis in bone tissue, but also is expressed by activated T cells, suggesting that osteoclastic bone resorption is influenced by T cells [8]. However, it is notable that most of the T-cell cytokines, including IFN-γ, IL-4 and IL-10, inhibit osteoclastogenesis [16]. Our laboratory identified IL-17-producing T-helper (Th17) cells as the exclusive osteoclastogenic T-cell subset [12]. IL-17 induces RANKL on osteoclastogenesis-supporting mesencymal cells such as osteoblasts. IL-17 also enhances local inflammation and increases the production of inflammatory cytokines, which further promote RANKL expression and activity (Fig. 1). Thus, the interactions of T cells and osteolcasts have a crucial role in the pathogenesis in bone loss in arthritis [14].

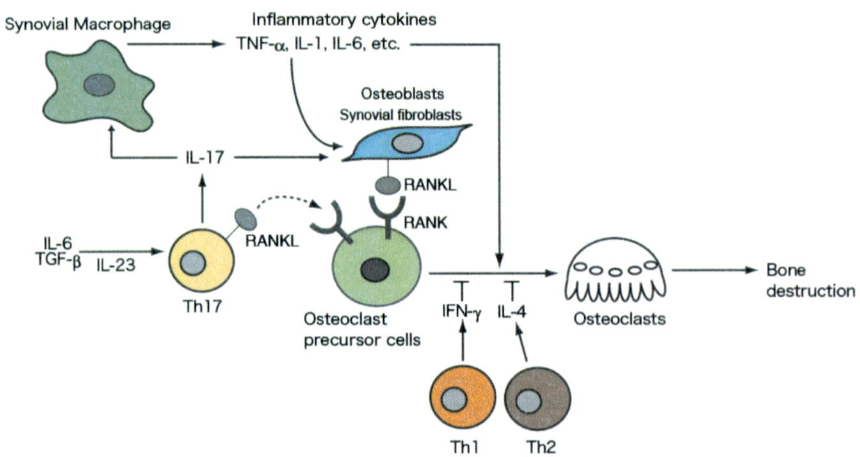

Fig. 1 Regulation of osteoclast differentiation by T cells in inflammation. T helper 17 (Th17) cells have stimulatory effects on osteoclastogenesis and play an important role in the pathogenesis of rheumatoid arthritis through IL-17, while Th1 and Th2 cells have inhibitory effects on osteoclastogenesis through IFN-γ and IL-4, respectively. IL-17 not only induces RANKL on osteoblasts/synovial fibroblasts of mesenchymal origin, but also activates local inflammation, leading to the upregulation of inflammatory cytokines such as TNF-α, IL-1 and IL-6. These cytokines further increase RANKL on mesenchymal cells and, in addition, act directly on the osteoclast precursor cells to enhance the RANKL activity

2 Transcriptional Machinery Activated by RANKL

RANKL activates the transcription-factor complex, activator protein 1 (AP1), through the induction of its component c-Fos [17]. The induction mechanism of c-Fos is not completely understood, but recent reports suggest that it is dependent on the activation of Ca^{2+}/calmodulin-dependent kinase (CaMK) IV and the cAMP response element-binding protein (CREB) [11], as well as the activation of NF-κB [18]. RANKL specifically and potently induces nuclear factor of activated T cells cytoplasmic 1 (NFATc1), the master regulator of osteoclast differentiation, and this induction is dependent on both the TRAF6 and c-Fos pathways [15]. The activation of NFAT is mediated by a specific phosphatase, calcineurin, which is activated by calcium–calmodulin signaling. The *NFATc1* promoter contains NFAT binding sites and NFATc1 specifically autoregulates its own promoter during osteoclastogenesis, thus effecting the robust induction of NFATc1 [2]. NFATc1 regulates a number of osteoclast-specific genes such as cathepsin K, tartrate-resistant acid phosphatase (TRAP), calcitonin receptor, OSCAR and β3 integrin in cooperation with other transcription factors such as AP1, PU.1, MITF and CREB [14].

NFATc1 was originally identified in T cells, but this transcription factor was subsequently revealed to play a crucial role in osteoclastogenesis. It is also worthy to note that calcium signaling and its downstream effector molecules are shared by osteoclasts and T cells.

3 Costimulatory Signals Along with RANK

Osteoclasts are formed when bone marrow cells are stimulated with RANKL and M-CSF in a culture system; therefore, it has been thought that RANK and c-Fms together transmit the signals sufficient for osteoclastogenesis. However, a novel type of receptor has been found to be highly expressed in osteoclasts. Osteoclast-associated receptor (OSCAR) was shown to be an immunoglobulin-like receptor involved in the cell-cell interaction between osteoblasts and osteoclasts [6]. Subsequent studies showed that OSCAR associates with an adaptor molecule, Fc receptor common γ subunit (FcRγ) [7], which harbors an immunoreceptor tyrosine-based activation motif (ITAM) critical for the activation of calcium signaling in immune cells. Another ITAM-harboring adaptor, DNAX activating protein 12 (DAP12), has been reported to be involved in the formation and function of osteoclasts [5]. Importantly, mice doubly deficient in FcRγ and DAP12 exhibit severe osteopetrosis, owing to a differentiation blockade of osteoclasts; demonstrating that the immunoglobulin-like receptors associated with FcRγ and DAP12 are essential for osteoclastogenesis [7, 9]. These receptors include OSCAR, triggering receptor expressed in myeloid cells (TREM)-2, signal-regulatory protein (SIRP) β1, and paired immunoglobulin-like receptor (PIR)-A, although the ligand and exact function of each of these receptors remain to be determined.

ITAM-mediated signals cooperate with RANK to stimulate calcium signaling through ITAM phosphorylation and the resulting activation of Syk and PLCγ. Therefore, these signals should properly be called costimulatory signals for RANK. Initially characterized in natural killer and myeloid cells, the immunoglobulin-like receptors associated with FcRγ or DAP12 are thus identified as previously unexpected, but nevertheless essential partners of RANK during osteoclastogenesis. It is not fully understood how RANK can specifically induce osteoclastogenesis in cooperation with ITAM signaling, but it is at least partially explained by the observation that phosphorylation of ITAM is upregulated by RANKL [7, 9]. In addition, RANKL stimulation results in an increased expression of immunoreceptors, such as OSCAR, thereby augmenting the ITAM signal.

4 Tec Family Kinases in B Cells and Osteoclasts

It is also conceivable that RANK activates an as-yet unknown pathway that specifically synergizes with or upregulates ITAM signaling. We have shown that Tec family tyrosine kinases such as Btk and Tec are activated by RANK and are involved in the phosphorylation of PLCγ [13]. An osteopetrotic phenotype in *Tec*$^{-/-}$*Btk*$^{-/-}$ mice revealed that these two kinases play an essential role in the regulation of osteoclast differentiation. Tec and Btk are known to have key roles in proximal BCR signaling, as shown by the *Tec*$^{-/-}$*Btk*$^{-/-}$ mice, in addition to the impaired antibody production and severe immunodeficiency in X-linked Bruton type agammagloburinemia (XLA) patients with a mutation in Btk. Our study established a further, crucial role of the Tec kinases in linking the RANK and ITAM signals. This study also identified an osteoclastogenic signaling complex composed of Tec kinases and scaffold proteins which introduces a new paradigm for the signal transduction mechanism of osteoclast differentiation: ITAM phosphorylation results in the recruitment of Syk, which phosphorylates adaptor proteins such as BLNK and SLP-76. These, in turn, function as scaffolds to recruit Tec kinases and PLCγ to the osteoclast signaling complex, so as to induce the maximal activation of calcium influx. The revised version of the signal transduction mechanism of RANKL is summarized in Fig. 2.

5 Cathepsin K in Dendritic Cells

Cathepsin K has been thought to be specifically expressed in osteoclasts and to play an essential role in the degradation of bone matrix such as type I collagen. We developed a new orally active cathepsin K inhibitor and examined the effect of the inhibitor in osteoporosis, as well as arthritis models. The results were unexpected: cathepsin K suppression leads to the reduction of inflammation in the latter model. Cathepsin K, despite being expressed at a low level in dendritic cells, plays an important role in the activation of Toll-like receptor (TLR) 9 signaling [1]. CpG DNA-induced production of cytokines such as IL-6 and IL-23 was found to be impaired in cathepsin K inhibitor-treated or cathepsin K-deficient dendritic cells.

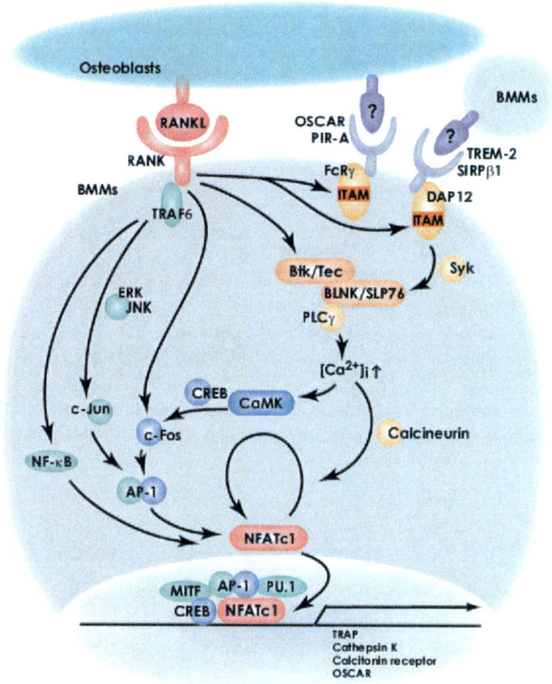

Fig. 2 Osteoclastogenic signal transduction. RANK activates the NF-κB and c-Fos pathways that stimulate the induction of NFATc1. Immunoglobulin-like receptors associate with ITAM-harboring adaptors, Fc receptor common γ subunit (FcRγ) and DNAX-activating protein 12 (DAP12). RANK and ITAM signaling cooperate to phosphorylate phospholipase Cγ (PLCγ) and activate calcium signaling, which is critical for the activation and autoamplification of NFATc1. Tec tyrosine kinases (Tec and Btk) activated by RANK are important for the formation of the osteoclastogenic signaling complex composed of Tec kinases, BLNK/SLP-76 (activated by ITAM-Syk) and PLCγ, which are essential for the efficient phosphorylation of PLCγ

The immune function of cathepsin K was further analyzed in experimental autoimmune encephalitis, and the severity of the disease was markedly suppressed in cathepsin K-deficient mice. The suppression of inflammation was associated with the reduced induction of Th17 cells, indicating that cathepsin K contributes to the autoimmune inflammation by inducing Th17 cells, possibly through cytokines such as IL-6 and IL-23 (Fig. 3).

The detailed mechanism remains to be elucidated, but cathepsin K is localized at the endosome, an organelle in which TLR9 recognizes the unmethylated DNA derived from pathogens. It is likely that cathepsin K is involved in the degradation of proteins that interfere with the association of pathogen DNA with TLR9. Thus, cathepsin K has been shown to play an unexpected role in the immune system, contributing to autoimmune inflammation. A cathepsin K inhibtor is expected to exert an affect not only on bone but also on the activation of dendritic cells.

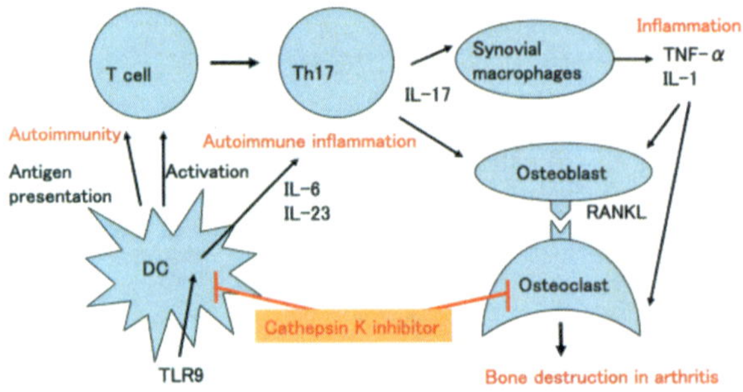

Fig. 3 A cathepsin K inhibitor suppresses both autoimmune inflammation and bone destruction. Cathepsin K is involved in the TLR9-mediated activation of dendritic cells, as well as osteoclastic bone resorption. Cathepsin K inhibition results in the reduced expression of inflammatory cytokines such as IL-6 and IL-23, which are important for the induction of Th17 cells. Thus, a cathepsin K inhibitor has dual benefits in the treatment of arthritis

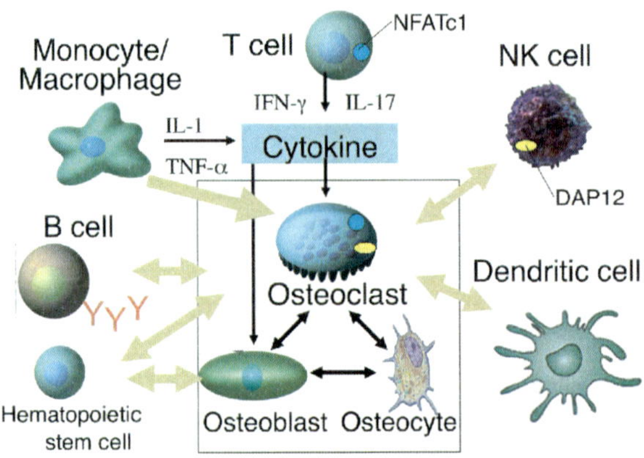

Fig. 4 Interaction of osteoclasts and various immune cells. The close relationship with osteoclasts has been extended from monocytes/macrophages and T cells to include other immune cells such as NK cells, B cells and dendritic cells (See the text for details)

6 Perspective

Recent studies indicated that osteoclasts interact and share molecules with most cells in the immune repertoire, including macrophages, T cells, NK cells, B cells and dendritic cells (Fig. 4). Further studies are absolutely required to understand fully the osteoclast-centered osteoimmunological interactions, which include new topics such as the transdifferentiation of dendritic cells to osteoclasts; the role of regulatory T cells in osteoclastogenesis, and the contribution of megakaryocytes to osteoclast biology.

The progress of osteoimmunology has been accelerated by analyzing mice deficient in immunomodulatory molecules. It is true that the molecules important in the immune system also play a distinct role in the skeletal system, but it is also becoming clear that molecules well known in the skeletal system play a crucial role in the immune system. These findings will lead to a better understanding of the pathogenesis of diseases affecting both systems, and will provide a molecular basis for novel therapeutic strategies. Osteoimmunology has emerged as a critical new interdisciplinary field as a result of transplanting techniques, methodologies and concepts from the field of the immune system to the field of bone biology. Now, it is time that the knowledge obtained in the analysis of bone has begun to provide unexpected information on the immune system as well.

Acknowledgments This work was supported in part by Grant-in-Aid for Creative Scientific Research from the Japan Society for the Promotion of Science (JSPS); grants for the Genome Network Project and the Global Center of Excellence (GCOE) Program from the Ministry of Education, Culture, Sports, Science, and Technology of Japan (MEXT); and grants from Tokyo Biochemical Research Foundation, Yokoyama Foundation for Clinical Pharmacology, the Japan Medical Association and Takeda Science Foundation.

References

1. Asagiri, M., Hirai, T., Kunigami, T. et al. (2008). Cathepsin K-dependent toll-like receptor 9 signaling revealed in experimental arthritis. *Science*, 319, 624–627
2. Asagiri, M., Sato, K., Usami, T. et al. (2005). Autoamplification of NFATc1 expression determines its essential role in bone homeostasis. *J Exp Med*, 202, 1261–1269
3. Dewhirst, F.E., Stashenko, P.P., Mole, J.E. et al. (1985). Purification and partial sequence of human osteoclast-activating factor: identity with interleukin 1β. *J Immunol* 135, 2562–2568
4. Horton, J.E., Raisz, L.G., Simmons, H.A. et al. (1972). Bone resorbing activity in supernatant fluid from cultured human peripheral blood leukocytes. *Science* 177, 793–795
5. Kaifu, T., Nakahara, J., Inui, M. et al. (2003). Osteopetrosis and thalamic hypomyelinosis with synaptic degeneration in DAP12-deficient mice. *J Clin Invest* 111, 323–332
6. Kim, N., Takami, M., Rho, J. et al. (2002). A novel member of the leukocyte receptor complex regulates osteoclast differentiation. *J Exp Med* 195, 201–209
7. Koga, T., Inui, M., Inoue, K. et al. (2004) Costimulatory signals mediated by the ITAM motif cooperate with RANKL for bone homeostasis. *Nature* 428, 758–763
8. Kong, Y.Y., Feige, U., Sarosi, I. et al. (1999). Activated T cells regulate bone loss and joint destruction in adjuvant arthritis through osteoprotegerin ligand. *Nature* 402, 304–309
9. Mocsai, A., Humphrey, M.B., Van Ziffle, J.A. et al. (2004). The immunomodulatory adapter proteins DAP12 and Fc receptor-chain (FcR) regulate development of functional osteoclasts through the Syk tyrosine kinase. *Proc Natl Acad Sci U S A* 101, 6158–6163
10. Mundy, G.R., Raisz, L.G., Cooper, R.A. et al. (1974). Evidence for the secretion of an osteoclast stimulating factor in myeloma. *N Engl J Med* 291, 1041–1046
11. Sato, K., Suematsu, A., Nakashima, T. et al. (2006a). Regulation of osteoclast differentiation and function by the CaMK-CREB pathway. *Nat Med* 12, 1410–1416
12. Sato, K., Suematsu, A., Okamoto, K. et al. (2006b). Th17 functions as an osteoclastogenic helper T cell subset that links T cell activation and bone destruction. *J Exp Med* 203, 2673–2682
13. Shinohara, M., Koga, T., Okamoto, K. et al. (2008). Tyrosine kinases Btk and Tec regulate osteoclast differentiation by linking RANK and ITAM signals. *Cell* 132, 794–806
14. Takayanagi, H. (2007). Osteoimmunology: shared mechanisms and crosstalk between the immune and bone systems. *Nat Rev Immunol* 7, 292–304
15. Takayanagi, H., Kim, S., Koga, T. et al. (2002). Induction and activation of the transcription factor NFATc1 (NFAT2) integrate RANKL signaling for terminal differentiation of osteoclasts. *Dev Cell* 3, 889–901

16. Takayanagi, H., Ogasawara, K., Hida, S. et al. (2000). T cell-mediated regulation of osteoclastogenesis by signalling cross-talk between RANKL and IFN-. *Nature* 408, 600–605
17. Wagner, E.F., & Eferl, R. (2005). Fos/AP-1 proteins in bone and the immune system. *Immunol Rev* 208, 126–140
18. Yamashita, T., Yao, Z., Li, F. et al. (2007). NF-κB p50 and p52 regulate receptor activator of NF-κB ligand (RANKL) and tumor necrosis factor-induced osteoclast precursor differentiation by activating c-Fos and NFATc1. *J Biol Chem* 282, 18245–18253

NFATc1 in Inflammatory and Musculoskeletal Conditions

Antonios O. Aliprantis and Laurie H. Glimcher

Abstract The nuclear factor of activated T-cells (NFAT) family of transcription factors specify developmental pathways and cell fate in vertebrates. NFATc1, in particular, is crucial to multiple seemingly unrelated biologic processes, including heart valve formation, T-cell activation, osteoclast development, and the mitigation of hair follicle stem cell proliferation. Here, we review how our recently generated NFATc1 conditional knockout mouse has contributed to our understanding of this transcription factor in inflammatory and musculoskeletal conditions and their treatment.

Keywords NFATc1 · Osteoclast · Hair follicle

1 Introduction

The calcium regulated NFAT family (NFATc1-c4) was originally identified 20 years ago as an activity that bound the IL-2 promoter in T-cells [35]. NFAT proteins are retained in the cytoplasm by hyperphosphorylation of an N-terminal regulatory domain. Signaling pathways, culminating in a sustained calcium influx, activate the phosphatase calcineurin, which dephosphorylates the regulatory domain and exposes a nuclear localization sequence. In the nucleus, NFAT proteins regulate gene transcription often in conjunction with other transcription factors, like AP-1 family members [2, 13]. Since the initial description of the NFAT family [35], our understanding of the numerous developmental and differentiation pathways that this vertebrate specific transcription factor family controls has grown immensely. NFATs have been implicated in immune cell activation, heart valve formation, cardiac hypertrophy, skeletal muscle development, vasculogenesis, cartilage formation,

A.O. Aliprantis (✉)
Department of Medicine, Harvard Medical School, Brigham and Women's Hospital and Department of Infectious Disease and Immunology, Harvard School of Public Health, Boston, MA, USA
e-mail: aaliprantis@partners.org

Y. Choi (ed.), *Osteoimmunology*, Advances in Experimental Medicine and Biology 658, DOI 10.1007/978-1-4419-1050-9_8, © Springer Science+Business Media, LLC 2010

neuron growth, and stem cell activation [8, 11, 12, 16, 28, 32, 30, 31, 47]. This insight has been driven, in large part, by the analysis of genetically engineered mice lacking individual or combinations of family members (reviewed most recently in [49]). Of the four family members, NFATc1 has been the most challenging to study in vivo, since *Nfatc1*$^{-/-}$ mice die in utero because of cardiac valve defects [27, 30]. Here, we describe how our recently generated *Nfatc1* conditional knockout mouse [1] has shed light on the role this transcription factor plays in the biology and treatment of inflammatory and musculoskeletal disease.

2 NFATc1 and Osteoclastogenesis

Bone mass is determined by the rates of bone formation by osteoblasts and bone resorption by osteoclasts [20]. Dysregulation of osteoclasts contributes to the pathogenesis of skeletal disorders, including osteoporosis and inflammatory arthritis, with excessive rates of resorption observed in both conditions [7, 29]. Osteoclasts are specialized, multinucleated giant cells that differentiate from myeloid precursors in the bone marrow (BM) in response to two cytokines produced by osteoblast lineage cells: Macrophage-Colony Stimulating Factor (M-CSF) and Receptor Activator of NF-κB Ligand (RANKL) [6, 40, 52]. M-CSF promotes the expansion and survival of pre-osteoclasts, while RANKL initiates the signaling cascades leading to osteoclast differentiation. A soluble decoy receptor for RANKL, called osteoprotegerin (OPG), which is also derived from osteoblasts, negatively regulates osteoclast differentiation in vivo [36]. Much has been learned about molecular mechanisms of bone destruction from genetic mutations identified in humans and mice that disrupt the formation or functional capabilities of the osteoclast, leading to a high bone mass phenotype, termed "osteopetrosis" [10, 18, 24, 26, 33, 37, 39, 41, 45, 46, 50].

The expression of NFATc1 is induced during osteoclastogenesis [23, 38] and this transcription factor can be identified at the promoters of many osteoclast specific genes [4, 9, 14, 21, 22, 37]. An examination of the role of NFATc1 in bone remodeling in vivo has been challenging since *Nfatc1*$^{-/-}$ embryos die of heart valve defects. Recently, fetal liver and blastocyst complementation approaches were used to show that cells lacking *Nfatc1* cannot rescue the osteopetrotic phenotype of the *c-fos* knockout mouse [4]. Furthermore, *Nfatc1*$^{-/-}$ mice rescued from embryonic lethality by intracardiac expression of NFATc1 display osteopetrosis. However, these mice die in the perinatal period, making an analysis of the role of NFATc1 in the adult skeleton is impossible [48]. Therefore, an examination of the role of NFATc1 in adult bone remodeling awaited definition; moreover, a tractable system to explore NFATc1 mediated gene expression was also lacking.

Recently, we generated a conditional knockout allele of *Nfatc1* (*Nfatc*^{fl}) and deleted it in mice at 10 days of age using *Mx1-Cre* [1]. *Nfatc1* deleted mice (*Nfatc1*$^{\Delta/\Delta}$) developed severe osteopetrosis characterized by increased bone mass, club shaped long bones, and a failure to degrade primary spongiosa at the growth plates, leading to a profound accumulation of calcified cartilage. Both in vivo and in vitro *Nfatc1* deficient cells failed to form osteoclasts. Accordingly, gene profiling of RANKL stimulated wild type and *Nfatc1* deficient osteoclast precursors, isolated to high purity by flow cytometry, and revealed this transcription factor as a master regulator of the osteoclast transcriptome. Interestingly, two classes of osteoclast-regulated genes were identified. Whereas some genes, such as *Mmp9, Acp5,* and *Ctsk* needed NFATc1 for optimal expression, others, like *Calcr*, *Itgb3,* and *Oscar* absolutely required NFATc1 for induction by RANKL. The reason why some genetic loci require NFATc1 more than others is likely to be complex and involve cross-talk between NFATc1 and other transcription factors that specify osteoclast differentiation like PU.1 and MITF [34] as well as molecules that drive chromatin remodeling. Our profiling experiments also revealed that in the absence of NFATc1, osteoclast precursors make OPG in response to RANKL. Accordingly, NFATc1 directly binds and represses the transcriptional activity of the OPG promoter [1]. The physiologic relevance of this observation remains to be determined, but the expression of OPG by multinucleated cells in giant cell tumors suggests this pathway may be activated in pathologic circumstances [5, 25]. Our data extend the requirement of NFATc1 to the remodeling of the growing and adult skeleton and highlight the central requirement of this transcription factor to osteoclast gene expression.

3 NFATc1 and Cherubism

Cherubism is a rare, pediatric, fibroinflammatory disease that causes disfiguring swelling of the face, bone loss in the jaw, and abnormalities in tooth development. Mutations in *SH3BP2*, which encodes a signaling adapter molecule, cause this disease [43]. A mouse model of cherubism has been recently described wherein the most common genetic alteration in cherubism patients, a missense mutation in exon 9 resulting in a P416R exchange, was knocked-in (*KI*) to the *Sh3bp2* locus [44]. Homozygous *KI/KI* mice develop systemic bone loss as well as inflammatory infiltrates in multiple organs, which are both dependent on the cytokine, TNF-α. Moreover, in vitro osteoclast precursors carrying mutant *Sh3bp2* alleles are hyperresponsive to RANKL, forming large and more numerous osteoclasts. The transcriptional mechanisms downstream of SH3BP2 leading to TNF-α driven inflammation and enhanced osteoclastogenesis are unclear.

Since NFATc1 is a master regulator of osteoclast differentiation [3, 4, 48] and NFAT family member have been implicated in inflammatory cascades, including TNFα production [15, 19, 42], *Nfatc1* was ablated in cherubism mice [1]. Homozygous *KI/KI* mice were crossed onto the *Nfatc1* conditional knockout strain ($Nfatc1^{fl/fl}$) developed in our laboratory. The conditional allele was deleted in the peri-natal period using *Mx1-Cre* to generate NFATc1 deficient mice ($Nfatc1^{\Delta/\Delta}$). *KI/KI* mice without NFATc1 were protected from the systemic bone loss characteristic of this strain. In contrast, these mice still developed mulitorgan inflammation in the liver, lung, spleen and stomach. Furthermore, osteoclastogenesis was completely abrogated in $Nfatc1^{\Delta/\Delta}$ *KI/+* and $Nfatc1^{\Delta/\Delta}$ *KI/KI* cells in vitro and in vivo, indicating the NFATc1 is downstream of SH3BP2 in the developing osteoclast [1].

Taken together, this study indicates that NFATc1 deficiency can uncouple bone loss from inflammatory cascades [1]. Thus, therapies targeting NFATc1 in cherubism patients may prevent osteolysis of the jaw but are unlikely to suppress the prominent inflammatory component of this disease and would need to be employed in conjunction with immunosuppressive modalities.

4 NFATc1 and Hair Follicle Growth: Explaining an Untoward Side Effect of Calcineurin Inhibitors

The calcineurin inhibitors, cyclosporine and FK506, are immunosuppressant drugs used to prevent organ transplant rejection and treat an array of inflammatory conditions such as ulcerative colitis and rheumatoid arthritis. Precocious hair growth is a side effect of these medications [51]. Recently, we showed that NFATc1 is specifically expressed in the stem cell niche of the hair follicle where it mitigates cell proliferation and, hence, entry of the follicle into anagen [17]. Accordingly, deletion of the conditional *Nfatc1* allele in the skin and its epidermal appendages using *K14-cre* resulted in accelerated hair growth. The capacity of NFATc1 to inhibit stem cell proliferation was linked to its ability to repress expression of cyclin-dependent kinase 4 (CDK4), an enzyme required for transition from G1 through S phase of the cell cycle [17]. These data explain why calcineurin inhibitors lead to hypertrichosis and identify a new pathway for the treatment of alopecia.

5 Conclusions and Future Directions

Thus far, we have used our *Nfatc1* conditional knockout strain to explore the role of this transcription factor in osteoclast gene expression, physiologic and pathologic bone remodeling, and the growth of hair follicle stem cells [1, 17]. The analysis of the *Nfatc1* conditional knockout strain in combination with other germline and conditional *Nfat* alleles should also prove insightful, as compound *Nfat* mutants often display more severe phenotypes than single mutants, reflecting compensatory functions among family member in different organ systems [49]. Thus, this strain will expand our understanding of how the NFAT family regulates vertebrate biology in health and disease.

References

1. Aliprantis, A.O., Ueki, Y., Sulyanto, R. et al. (2008). NFATc1 represses osteoprotegerin during osteoclastogenesis and dissociates systemic osteopenia from inflammation in cherubism. *J Clin Invest*, 118(11):377–389.

2. Arron, J.R., Winslow, M.M., Polleri, A. et al. (2006). NFAT dysregulation by increased dosage of DSCR1 and DYRK1A on chromosome 21. *Nature*, 441:595–600.
3. Asagiri, M., & Takayanagi, H. (2007). The molecular understanding of osteoclast differentiation. *Bone*, 40:251–264.
4. Asagiri, M., Sato, K., Usami, T. et al. (2005). Autoamplification of NFATc1 expression determines its essential role in bone homeostasis. *J Exp Med*, 202:1261–1269.
5. Atkins, G.J., Bouralexis, S., Haynes, D.R. et al. (2001). Osteoprotegerin inhibits osteoclast formation and bone resorbing activity in giant cell tumors of bone. *Bone*, 28:370–377.
6. Boyle, W.J., Simonet, W.S., & Lacey, D.L. (2003). Osteoclast differentiation and activation. *Nature*, 423:337–342.
7. Carlsten, H. (2005). Immune responses and bone loss: the estrogen connection. *Immunol Rev*, 208:194–206.
8. Chang, C.P., Neilson, J.R., Bayle, J.H. et al. (2004). A field of myocardial-endocardial NFAT signaling underlies heart valve morphogenesis. *Cell*, 118:649–663.
9. Crotti, T.N., Flannery, M., Walsh, N.C. et al. (2006). NFATc1 regulation of the human beta3 integrin promoter in osteoclast differentiation. *Gene*, 372:92–102.
10. Fuller, K., Wong, B., Fox, S. et al. (1998). TRANCE is necessary and sufficient for osteoblast-mediated activation of bone resorption in osteoclasts. *J.Exp.Med*, 188:997–1001.
11. Graef, I.A., Chen, F., & Crabtree, G.R. (2001). NFAT signaling in vertebrate development. *Curr Opin Genet Dev*, 11:505–512.
12. Graef, I.A., Wang, F., Charron, F. et al. (2003). Neurotrophins and netrins require calcineurin/NFAT signaling to stimulate outgrowth of embryonic axons. *Cell*, 113:657–670.
13. Gwack, Y., Sharma, S., Nardone, J. et al. (2006). A genome-wide Drosophila RNAi screen identifies DYRK-family kinases as regulators of NFAT. *Nature*, 441:582–583.
14. Hirotani, H., Tuohy, N.A., Woo, J.T. et al. (2004). The calcineurin/nuclear factor of activated T cells signaling pathway regulates osteoclastogenesis in RAW264.7 cells. *J Biol Chem*, 279:13984–13992.
15. Hodge, M.R., Ranger, A.M., Charles de la Brousse, F. et al. (1996). Hyperproliferation and dysregulation of IL-4 expression in NF-ATp-deficient mice. *Immunity*, 4:397–405.
16. Hogan, P.G., Chen, L., Nardone, J. et al. (2003). Transcriptional regulation by calcium, calcineurin, and NFAT. *Genes Dev*, 17:2205–2232.
17. Horsley, V., Aliprantis, A.O., Polak, L. et al. (2008). NFATc1 balances quiescence and proliferation of skin stem cells. *Cell*, 132:299–310.
18. Jimi, E., & Ghosh, S. (2005). Role of nuclear factor-kappaB in the immune system and bone. *Immunol Rev*, 208:80–87.
19. Kaminuma, O., Kitamura, F., Kitamura, N. et al. (2008). Differential contribution of NFATc2 and NFATc1 to TNF-alpha gene expression in T cells. *J Immunol*, 180:319–326.
20. Karsenty, G., & Wagner, E.F. (2002). Reaching a genetic and molecular understanding of skeletal development. *Dev Cell*, 2:389–406.
21. Kim, K., Kim, J.H., Lee, J. et al. (2005a). Nuclear factor of activated T cells c1 induces osteoclast-associated receptor gene expression during tumor necrosis factor-related activation-induced cytokine-mediated osteoclastogenesis. *J Biol Chem*, 280:35209–35216.
22. Kim, Y., Sato, K., Asagiri, M. et al. (2005b). Contribution of nuclear factor of activated T cells c1 to the transcriptional control of immunoreceptor osteoclast-associated receptor but not triggering receptor expressed by myeloid cells-2 during osteoclastogenesis. *J Biol Chem*, 280:32905–32913.
23. Koga, T., Inui, M., Inoue, K. et al. (2004). Costimulatory signals mediated by the ITAM motif cooperate with RANKL for bone homeostasis. *Nature*, 428:758–763.
24. Kong, Y.Y., Yoshida, H., Sarosi, I. et al. (1999). OPGL is a key regulator of osteoclastogenesis, lymphocyte development and lymph-node organogenesis. *Nature*, 397:315–323.
25. Liu, B., Yu, S.F., & Li, T.J. (2003). Multinucleated giant cells in various forms of giant cell containing lesions of the jaws express features of osteoclasts. *J Oral Pathol Med*, 32:367–375.
26. Lorenzo, J., & Choi, Y. (2005). Osteoimmunology. *Immunol Rev*, 208:5–6.

27. Luisde la Pompa, J., Timmerman, L.A., Takimoto, H. et al. (1998). Role of the NF-ATc transcription factor in morphogenesis of cardiac valves and septum. *Nature*, 392:182–186.
28. Peng, S.L., Gerth, A.J., Ranger, A.M. et al. (2001). NFATc1 and NFATc2 together control both T and B cell activation and differentiation. *Immunity*, 14:13–20.
29. Phillips, K., Aliprantis, A., & Coblyn, J. (2006). Strategies for the prevention and treatment of osteoporosis in patients with rheumatoid arthritis. *Drugs Aging*, 23:773–779.
30. Ranger, A.M., Grusby, M.J., Hodge, M.R. et al. (1998a). The transcription factor NF-ATc is essential for cardiac valve formation. *Nature*, 392:186–190.
31. Ranger, A.M., Hodge, M.R., Gravallese, E.M. et al. (1998b). Delayed lymphoid repopulation with defects in IL-4-driven responses produced by inactivation of NFATc. *Immunity*, 8:125–134.
32. Ranger, A.M., Gerstenfeld, L.C., Wang, J. et al. (2000). The transcription factor NFATp is a repressor of chondrogenesis. *J Exp Med*, 191:9–21.
33. Rho, J., Takami, M., & Choi, Y. (2004). Osteoimmunology: interactions of the immune and skeletal systems. *Mol Cells*, 17:1–9.
34. Sharma, S.M., Bronisz, A., Hu, R. et al. (2007). MITF and PU.1 recruit p38 MAPK and NFATc1 to target genes during osteoclast differentiation. *J Biol Chem*, 282:15921–15929.
35. Shaw, J., Utz, P., Durand, D. et al. (1988). Identification of a putative regulator of early T cell activation genes. *Science*, 241:202–205.
36. Simonet, W.S., Lacey, D.L., Dunstan, C.R. et al. (1997). Osteoprotegerin: a novel secreted protein involved in the regulation of bone density. *Cell*, 89:309–319.
37. Takayanagi, H., Sato, K., Takaoka, A. et al. (2005). Interplay between interferon and other cytokine systems in bone metabolism. *Immunol Rev*, 208:181–193.
38. Takayanagi, H., Kim, S., Koga, T. et al. (2002). Induction and activation of the transcription factor NFATc1 (NFAT2) integrate RANKL signaling in terminal differentiation of osteoclasts. *Dev Cell*, 3:889–901.
39. Tanaka, S., Nakamura, K., Takahasi, N. et al. (2005). Role of RANKL in physiological and pathological bone resorption and therapeutics targeting the RANKL-RANK signaling system. *Immunol Rev*, 208:30–49.
40. Tanaka, S., Takahashi, N., Udagawa, N. et al. (1993). Macrophage colony-stimulating factor is indispensable for both proliferation and differentiation of osteoclast progenitors. *J Clin Invest*, 91:257–263.
41. Tolar, J., Teitelbaum, S.L., & Orchard, P.J. (2004). Osteopetrosis. *N Engl J Med*, 351: 2839–2849.
42. Tsytsykova, A.V., & Goldfeld, A.E. (2000). Nuclear Factor of Activated T Cells Transcription Factor NFATp Controls Superantigen-induced Lethal Shock. *J Exp Med*, 192:581–586.
43. Ueki, Y., Tiziani, V., Santanna, C. et al. (2001). Mutations in the gene encoding c-Abl-binding protein SH3BP2 cause cherubism. *Nat Genet*, 28:125–126.
44. Ueki, Y., Lin, C.Y., Senoo, M. et al. (2007). Increased myeloid cell responses to M-CSF and RANKL cause bone loss and inflammation in SH3BP2 "cherubism" mice. *Cell*, 128:71–83.
45. Wada, T., Nakashima, T., Hiroshi, N. et al. (2006). RANKL-RANK signaling in osteoclastogenesis and bone disease. *Trends Mol Med*, 12:17–25.
46. Wagner, E.F., & Eferl, R. (2005). Fos/AP-1 proteins in bone and the immune system. *Immunol Rev*, 208:126–140.
47. Wein, M.N., Jones, D.C., & Glimcher, L.H. (2005). Turning down the system: counter-regulatory mechanisms in bone and adaptive immunity. *Immunol Rev*, 208:66–79.
48. Winslow, M.M., Pan, M., Starbuck, M. et al. (2006). Calcineurin/NFAT signaling in osteoblasts regulates bone mass. *Dev Cell*, 10:771–782.
49. Wu, H., Peisley, A., Graef, I.A. et al. (2007). NFAT signaling and the invention of vertebrates. *Trends Cell Biol*, 17:251–260.
50. Xing, L., Schwarz, E.M., & Boyce, B.F. (2005). Osteoclast precursors, RANKL/RANK, and immunology. *Immunol Rev*, 208:19–29.

51. Yamamoto, S., & Kato, R. (1994). Hair growth-stimulating effects of cyclosporin A and FK506, potent immunosuppressants. *J Dermatol Sci*, 7(Suppl):S47–S54.
52. Yasuda, H., Shima, N., Nakagawa, N. et al. (1998). Osteoclast differentiation factor is a ligand for osteoprotegerin/osteoclastogenesis-inhibitory factor and is identical to TRANCE/RANKL. *Proc Natl Acad Sci USA*, 95:3597–3602.

Novel Functions of RANK(L) Signaling in the Immune System

Andreas Leibbrandt and Josef M. Penninger

Abstract The TNF family members RANKL and its receptor RANK have initially been described as factors expressed on T cells and dendritic cells (DCs), respectively, and have been shown to augment the ability of DCs to stimulate naive T cell proliferation and enhance DC survival. Since another, yet soluble receptor for RANKL, namely OPG, was initially characterized as a factor inhibiting osteoclast development and bone resorption, it was somewhat enigmatic at first why one and the same genes would be essential both for the immune system and bone development – two processes that on first sight do not have much in common. However, in a series of experiments it was conclusively shown that RANKL-expressing T cells can also activate RANK-expressing osteoclasts, and thereby in principal mimicking RANKL-expressing osteoblasts. These findings lead to a paradigm shift and helped to coin the term osteoimmunology in order to account for the crosstalk of immune cells and bone. More importantly was that these findings also provided a rationale for the bone loss observed in patients with a chronically activated immune system such as in rheumatoid arthritis, leukemias, or the like, arguing that T cells, which were activated during the course of the disease to fight it off, also express RANKL, which induces osteoclastogenesis and thereby shifts the intricate balance of bone deposition and resorption in favor of the latter. Through knockout mice it became also clear that the RANKL-RANK-OPG system is involved in other processes such as in controlling autoimmunity or immune responses in the skin. We will briefly summarize the role of RANK(L) signaling in the immune system before we discuss some of the recent data we and others have obtained on the role of RANK(L) in controlling autoimmunity and immune responses in the skin.

Keywords RANK · RANKL · OPG · Osteoimmunology · Osteoclast · Osteoclastogenesis · Immune system · Rheumatoid arthritis · Osteoporosis · Autoimmunity

---

J.M. Penninger (✉)
IMBA, Institute of Molecular Biotechnology of the Austrian Academy of Sciences, Dr. Bohr-Gasse 3, A-1030 Vienna, Austria
e-mail: josef.penninger@imba.oeaw.ac.at

Y. Choi (ed.), *Osteoimmunology*, Advances in Experimental Medicine and Biology 658, DOI 10.1007/978-1-4419-1050-9_9,

Abbreviations

TNF	Tumor Necrosis Factor
RANKL	Receptor Activator of Nuclear Factor-κB (NF-κB) Ligand
RANK	Receptor Activator of Nuclear Factor-κB (NF-κB)
OPG	Osteoprotegerin
DCs	Dendritic Cells
LT	Lymphotoxin
PPs	Peyer's Patches
LNs	Lymph Nodes
ALPS	Autoimmune Lymphoproliferative Syndrome
cTEC	Cortical Thymic Epithelial Cell
mTEC	Medullary Thymic Epithelial Cell
TRA	Tissue Restricted Antigen
AIRE	Autoimmune Regulator
LTi	Lymphoid Tissue Inducer
Tregs	eRgulatory T Cells
LCs	Langerhans Cells
RA	Rheumatoid Arthritis
BMD	Bone Mineral Density

1 RANK(L) Signaling in the Immune System

1.1 Lymph Node Development

Surprisingly, the detailed analysis of both Rankl$^{-/-}$ and Rank$^{-/-}$ revealed an entirely unexpected phenotype – the complete absence of all lymph nodes [1–3]. Other TNF family molecules have also been implicated in the development and organization of secondary lymphoid tissues, such as lymphotoxin-α (LTα) [4], LTβ [5, 6], TNFR1 [7], or LTβ receptor (LTβR) [8]. It is important to note that LTα knockout mice lack all lymph nodes (LNs), Peyer's patches (PPs), and follicular dendritic cells; they also show altered splenic architecture [6]. Since LTα not only forms (soluble) homotrimers but also heterotrimers with LTβ, the disruption of LTα alone and the effect on peripheral lymphoid organs cannot rule out a critical involvement of LTβ. Thus, LTβ knockout mice have been generated, which show similar defects in the organogenesis of the lymphoid system as LTα: they lack PPs, peripheral lymph nodes, splenic germinal centers, and follicular dendritic cells, but most LTβ$^{-/-}$ mice still retain mesenteric and cervical lymph nodes, suggesting an LTβ-independent function of LTα in the development of these LNs [5, 6].

However, LTβR knockout mice lack PPs, colon-associated lymphoid tissues, and importantly all lymph nodes [8]. The lack of lymph nodes in LTα mutant mice is best explained by defective LTβR signaling that is further strengthened by the observation that LTβR$^{-/-}$ mice also lack all lymph nodes. The observation of some lymph nodes in LTβ-deficient mice would, however, argue that another cytokine co-operates with LTα in the development of mesenteric and cervical lymph nodes. Since a common denominator of mice with disrupted LT signaling is the lack of PPs, it was initially assumed that PP and LN development are alwaysgenetically linked. However, the analysis of Rankl- and Rank-deficient mice showed that these mutant mice display intact splenic architecture and lack all lymph nodes but develop PPs normally, providingevidence, for the first time, that development of LNs and PPs can be genetically uncoupled [1–3]. In patients with an osteoclast-poor form of autosomal recessive osteopetrosis (ARO), various mutations in RANKL have been identified as the cause of the disease [9]. These patients presented with no palpable lymph nodes, suggesting that RANKL-RANK signaling also controls lymph node formation in humans [9].

Several distinct cell lineages such as fibroblasts, macrophages, reticular cells, and endothelial cells are required for primordial lymph node development [10]. These primordial lymph nodes are subsequently seeded by T and B cells and $CD4^{+}CD3^{-}LT\beta,^{+}$ cells that differentiate into NK cells, antigen presenting cells, and follicular cells to form mature compact nodes [11]. Of noted importance is that RANKL- and RANK-expressing cells are present in the "mature" lymph nodes, located mainly in the cortical areas adjacent to subcapsular sinuses [12]. Since RANK and RANKL are also expressed in the spleen and PPs, restricted RANKL-RANK expression cannot account for the selective lack of lymph nodes. Moreover, transfer experiments of $Rankl^{-/-}$ fetal liver cells into $Rag1^{-/-}$ mice indicated that the lack of lymph nodes in $Rankl^{-/-}$ is not due to a cellular homing defect of $Rankl^{-/-}$ lymphocytes; transfer of normal bone marrow cells into newborn Rankl knockout mice did also not rescue lymph node formation [1]. Another study could show that defective LN development in $Rankl^{-/-}$ mice correlates with a significant reduction of $\alpha_4\beta_7{}^{+}CD45^{+}CD4^{+}CD3^{-}$ cells in developing LN anlagen and their failure to cluster.

In $LT\alpha^{-/-}$ mice, these hematolymphoid precursor cells also showed impaired colonization and cluster formation in the LN anlage. Transgenic overexpression of RANKL could not restore LN development in $LT\alpha^{-/-}$ mice, showing that $LT\alpha_1\beta_2$ expression is required on $\alpha_4\beta_7{}^{+}CD45^{+}CD4^{+}CD3^{-}$ cells [13]. Thus, these results indicate that both RANKL and lymphotoxin ligands regulate the LN genesis by controlling the colonization and cluster formation of $\alpha_4\beta_7{}^{+}CD45^{+}CD4^{+}CD3^{-}$ cells during LN development. Since the defects in the RANKL/RANK or LT/LTβR system did not affect the initiation of colonization by $\alpha_4\beta_7{}^{+}CD45^{+}CD4^{+}CD3^{-}$ cells, the results would also suggest that neither RANKL nor LT are intrinsically required for the initiation of the formation of LN anlage, but rather they are required for further development of the LN anlage [13]. Although some light has been shed on LN development over the last years, the exact cellular and molecular mechanism of RANKL-RANK-regulated LN organogenesis as well as the link between LN and Peyer's patch formation still await resolution.

1.2 Dendritic Cells

Dendritic cells (DCs) are cells specialized to capture and process antigens. In most tissues, DCs are present in an immature state unable to stimulate T cells. Contact with antigen leads to their maturation in response to inflammatory stimuli. Mature DCs that have captured antigens then migrate to T cell zones of secondary lymphoid organs by afferent lymphatics and present antigen to antigen-specific T cells. The T cell areas of secondary lymphoid organs represent the microenvironment to allow interactions between DCs, T cells, and B cells to initiate adaptive immune responses [14]. Antigen-bearing DCs are in direct contact with naive antigen-specific T cells within the T cell areas of lymph nodes and after interaction with T cells these DCs are rapidly eliminated [15]. For DC elimination to occur, activated T cells induce apoptosis of DCs by producing the TNF family molecules TRAIL, FasL, and TNF-α. The relevance of DC death after antigen presentation could be shown in patients with autoimmune lymphoproliferative syndrome (ALPS), an inherited disease of lymphocyte homeostasis and defective apoptosis. ALPS patients manifest lymphocytosis and autoimmune disorders. In the type II form of ALPS, DCs accumulation and prolonged DC survival were shown to be due to a caspase 10 mutation that rendered DCs resistant to TRAIL-induced cell death [16]. These results indicated that mature DCs presenting antigens to T cells have to be effectively eliminated in order to avoid excessive immune responses. The life span of DCs might thus be an important checkpoint to control for the induction of tolerance, priming, and chronic inflammation.

RANKL is not expressed on resting $CD4^+$ or $CD8^+$ T cells, but 4 h after anti-CD3/CD28 stimulation, surface RANKL is detected on $CD4^+$ T cells with a peak at 48 h and sustained high levels until at least 96 h. RANKL expression on $CD8^+$ T cells follows similar kinetics upon stimulation but generally with lower levels of surface RANKL than $CD4^+$ T cells [17]. RANK surface expression can be detected on DCs from various sources – on mature bone marrow-derived DCs, freshly isolated lymph node DCs, or freshly isolated splenic Dcs, which can even be increased by overnight culture in vitro, but RANK cannot be detected on freshly isolated LN B cells, LN-derived T cells, thymocytes, or peritoneal macrophages [18]. In fact, interactions between RANKL on activated T cells and RANK on DCs have been shown to mediate DC survival via Bcl-x_L induction and upregulation of the costimulatory molecule CD40 on DCs [18–20]. Since RANKL activates the anti-apoptotic Ser/Thr kinase Akt/PKB, NF-κB and ERK through a signaling complex involving TRAF6 and c-Src on mature DCs and osteoclasts [21], activation of anti-apoptotic molecules seems to be at least partially responsible for RANKL-mediated DC survival [22].

Moreover, it has been shown that treatment of antigen-pulsed mature DCs with soluble RANKL in vitro before immunization enhances the number and persistence of antigen-presenting DCs in the draining LNs in vivo, and RANKL treatment also increased antigen-specific primary T cell responses. Interestingly, significant memory responses were observed only in mice injected with RANKL-treated DCs [20]. Since RANKL can induce multiple cytokines in DCs like IL-1, IL-6, IL-12 and IL-15 [17, 23], it could well be that these increased primary and memory T cells responses following vaccination with RANKL-treated DCs are due to enhanced/altered cytokine production. Alternatively, the enhanced T cell responses by RANKL might also be due to increased numbers of antigen-pulsed DCs found in draining LNs [20]. CD40L, a TNF family member closely related to RANKL, shows functional similarity to RANKL, in that both CD40L and RANKL are expressed on activated T cells and enhance the activation and survival of DCs [18, 19]. However, in contrast to the CD40L-CD40 system, RANKL-RANK signaling from T cells to DCs does not alter the expression of cell surface molecules such as MHC class II, CD80, CD86, and CD54. Whereas CD40L is primarily expressed on activated $CD4^+$ T cells, RANKL is expressed on both activated $CD4^+$ and $CD8^+$ T cells [17, 24]. However, the kinetics of RANKL and CD40L expression are somewhat different: maximal RANKL levels following initial T cell activation peaked around 48 h and remained high at least until 96 h, while CD40L reaches maximal levels between 6 and 8 h and is downregulated to resting levels between 24 and 48 h [17, 25]. Thus, CD40L-CD40 interactions might primarily control the initial priming stage whereas RANKL-RANK might act at later time points than CD40L during the immune response.

Although CD40L-CD40 interactions are crucial for the generation of antigen-specific T cell responses in vivo [26], the finding that CD40- and CD40L-deficient mice can still mount protective $CD4^+$ T cell responses upon viral challenge in vivo [27] would suggest that some pathogens can activate $CD4^+$ T cells independent of CD40L signaling. The factors that might mediate the CD40(L)-independent $CD4^+$ T cell priming could be RANKL and RANK: inhibition of RANKL in vivo by soluble RANK-Fc did not block the priming of lymphocytic choriomeningitis virus-specific T cells but impaired proliferation of $CD4^+$ T cells to the viral antigen at later time points after infection and was especially apparent in the absence of CD40-expression [23]. Thus, at later stages of the immune response, RANKL can regulate CD40L-independent activation of $CD4^+$ T helper cells. These observations suggested that although CD40L and RANKL have functional similarity and might cooperate, RANKL and CD40L might also have fundamentally different functions in the control of immune responses: CD40L might regulate T and B cell responses while RANKL appears to have a role in memory T cell responses.

The decoy receptor for RANKL, OPG, can also be found on the cell surface of DCs and was furthermore shown to bind to the TNF-family molecule TRAIL, which is produced by activated T cells to induce apoptosis of DCs [28]. Thus, it seems as if the balance between RANKL and TRAIL – both produced from activated T cells – could influence DC survival and OPG might modulate that regulatory loop. However, since the binding affinity of OPG to TRAIL is rather low ($\sim$10,000 times less binding to TRAIL than to RANKL) [12] , it is still unclear whether OPG-TRAIL interactions have any functional relevance in vivo. Based on these studies, various groups are currently trying to control the DC fate via RANKL-RANK and OPG to modulate in vivo DC survival and to enhance the efficacy of DC-based vaccinations for anti-tumor therapy or the treatment of autoimmune diseases. However, in the final analyses of all the published genetic and functional studies on RANKL, RANK, and OPG, it appears that although these molecules can influence some aspects of lymphocyte and DC functions, none of these molecules plays an essential role in T, B, or dendritic cells that cannot be compensated for by other molecules such as CD40L/CD40. Since we and others have shown that expression of RANKL and OPG molecules can be controlled by sex hormones [29], RANKL-RANK might control gender specific differences in immunity and could be involved in the higher incidence of autoimmune diseases like arthritis in women.

1.3 RANK(L) Signaling in the Thymus

In our initial RANKL knock-out paper, we had actually observed altered T cell development and expression of RANKL on primarily $CD4^{-}8^{-}$ thymocyte progenitors. However, since RANK mutant mice exhibited apparently normal thymocyte development it was unclear whether this effect was secondary to other in vivo phenotypes in our mice and whether RANKL-RANK might indeed have some primary functions in the thymus. Interestingly, a recent study provided evidence for such a primary function of RANKL-RANK in the thymus; surprisingly, RANKL-RANK controls the development of $AIRE^{+}$ thymic epithelial cells.

Cortical thymic epithelial cells (cTECs) positively select thymocytes, which subsequently migrate to the medulla where thymocytes interact with medullary thymic epithelial cells (mTECs), expressing costimulatory molecules and self-tissue-restricted antigens (TRAs) [30]. TRA expression has been shown to be regulated in part by the transcription factor AIRE (autoimmune regulator). AIRE mutations lead to multiorgan autoimmune disease in humans and autoimmunity in AIRE gene targeted mice. This phenotype could be mapped to a subtype of thymic medullary epithelial cells, thereby demonstrating the importance of $Aire^{+}$ mTECs in maintaining self-tolerance [31]. Despite the identification of a common progenitor of cTECs and mTECs [32], the developmental pathway leading to $Aire^{+}$ mTECs was not entirely clear.

However, it has recently been demonstrated that an intrathymic $CD4^{+}3^{-}$ lymphoid tissue inducer (LTi) cell population expresses RANKL and that RANKL signals from thymic LTi cells to $CD80^{-}Aire^{-}$ mTECs are required for their development into $CD80^{+}Aire^{+}$ mTECs [33]. In line with these data, Aire expression is absent in thymi from $Rank^{-/-}$ mice and transplantation of $Rank^{-/-}$ thymic tissue under the kidney capsule of nude mice resulted in inflammatory infiltrates in liver and autoantibodies to several tissues, which parallels the phenotype observed after transplantation of $Aire^{-/-}$ thymic stroma [31, 33]. These findings suggest a key role of RANK signals in the regulation of central tolerance for which LTα-LTβR signals have also been shown to be important [34, 35]. However, $Aire^{+}$ mTECs can develop in the absence of LTβR and LTα [33, 35], suggesting that some aspects of mTEC development and organization may involve both RANK and LTβR signals. Moreover, in the absence of Traf6, the key downstream signaling adaptor of RANK, $Aire^{+}$ mTECs also do not develop leading to the onset of organ-specific autoimmunity [36] . Thus, RANKL expressed by inducer cells, and possibly other thymic cell types, can activate RANK on thymic epithelial progenitor cells to develop into $AIRE^{+}$ thymic medullary epithelial cells.

In three recent studies, further evidence for a crucial role of RANKL-RANK interactions in mediating mTEC development was presented. In one study, the authors could demonstrate that mTEC development required both RANK and CD40 signaling [37]. Interestingly, RANK signaling was essential only for mTEC development during embryogenesis, since mTECs failed to develop properly in embryonic thymus while mTECs – albeit at reduced numbers – were detected in postnatal mice even in the complete absence of RANKL. The signal capable of taking over for RANKL in postnatal thymus was shown to be CD40L, since development of mTECs was almost completely abolished in CD40-RANKL double mutant mice. Moreover, CD40L- or RANKL-mediated mTEC development was also shown to be dependent on Traf6, NIK, and IKKβ, respectively. These results show that developmental-stage-dependent co-operation between RANK and CD40 promotes mTEC development, thereby establishing self-tolerance [37].

In the second study, the authors could show that positively selected thymocytes express RANKL and expand mTEC cellularity to form the mature thymic medulla [38]. This effect, importantly, effect depended on interaction with RANK and OPG expressed by mTECs, since neutralization of RANKL by RANK-Fc expression perturbed mTEC cellularity and RANKL expression in mice deficient for positive selection restored thymic medulla. These results suggest that RANKL produced by positively selected thymocytes can also foster thymic medulla formation, thereby establishing central tolerance [38].

In order to address the role of single-positive $CD4^+$ and $CD8^+$ thymocytes in the process of postnatal mTEC development, a recent third study went on to show that although either $CD4^+$ or $CD8^+$ thymocytes were sufficient to sustain formation of a well-defined medulla expansion of the mature mTEC population required autoantigen-specific interactions between positively selected $CD4^+$ thymocytes bearing autoreactive T cell receptor (TCR) and mTECs displaying cognate self-peptide-MHC class II complexes. These interactions were shown to involve the engagement of CD40 on mTECs by CD40L induced on the positively selected $CD4^+$ thymocytes. This antigen-specific TCR-MHC class II-mediated crosstalk between $CD4^+$ thymocytes and mTECs thus defines a unique checkpoint in thymic stromal development that is pivotal for generating a mature mTEC population competent for ensuring central T cell tolerance [39].

1.4 Extramedullary Haematopoiesis and B Cell Phenotypes

In addition to T cells, $Rankl^{-/-}$ and $Rank^{-/-}$ mice have reduced numbers of mature $B220^+IgD^+$ and $B220^+IgM^+$ B cells in the spleen and lymph nodes and slightly disorganized B cell areas in primary splenic follicles [1, 2, 46]. Since $Rankl^{-/-}$ and $Rank^{-/-}$ mice have no bone marrow cavities, the reduced cellularity of B cells could be due to an altered microenvironment or due to changes in the composition of stromal cells outside the bone marrow cavity that affect B cell differentiation. For example, $Rankl^{-/-}$ mice form an ectopically organized extramedullary haematopoietic tissue localized at the outer surfaces of vertebral bodies [1]. This tissue exhibits morphological and phenotypic features characteristic of haematopoiesis and proliferating precursor cells. Whether these haematopoietic islands in Rankl$^{-/-}$ mice represent a defect in the homing of precursors during the switch from hepatic to bone marrow haematopoiesis or an event secondary to osteopetrosis, which interferes with the seeding of bone marrow cavities, remains to be determined. In fetal liver cell chimeras, RANKL was found to regulate early B cell differentiation from the B220^{+}CD43^{+}25^{-} pro-B cell to the B220^{+}CD43^{-}25^{+} pre-B cell stage of development, indicating suggesting the TNF-family cytokine RANKL is indeed a regulator of early B lymphocyte development [1]. Evidence in an Opg mutant mouse strain confirmed the notion that the interplay RANKL-RANK and the molecular decoy receptor OPG may regulate the development and possibly the function of B lymphocytes [47]. Ex vivo, Opg$^{-/-}$ pro-B cells have enhanced proliferation to IL-7 and type 1 transitional B cells accumulate in the spleens of Opg$^{-/-}$ mice. Thus, loss of OPG may control B cell maturation. Moreover, it should be noted that OPG is a CD40-regulated gene in B cells and dendritic cells and that prostaglandin E2 treatment can increase the amount of RANKL messenger RNA in B220^{+} B cells in an estrogen-dependent manner [47, 48].

2 RANKL-RANK Signaling Can Mediate UV-Induced Immunosuppression

Recently, together with the group of Stefan Beissert, we could show that RANKL expression in the skin can control the number of regulatory T cells (Tregs). Tregs, in particular CD4^{+}25^{+} Tregs expressing the transcription factor Foxp3, are a functionally distinct T cell subpopulation. Tregs maintain immunological self-tolerance and suppress excessive immune responses to self-antigens such as those in autoimmune diseases or allergies [40]. Despite the importance of DCs in inducing immunity to infections, it has been shown that DCs can also induce expansion of CD4^{+}25^{+} Tregs and thereby induce T cell tolerance [41]. Given that activation of epidermal Langerhans cells (LCs; dendritic cells of the skin) by CD40L, a TNF family member closely related to RANKL, can induce severe systemic autoimmunity [42] and the importance of RANKL-RANK signals in T cell-DC interactions, we speculated that RANK signaling might also be important for immune homeostasis in the skin. Moreover, skin is the main site for sun light-induced Vitamin D3 production; Vitamin D3 is one of the key triggers of RANKL expression during osteoclastogenesis.

RANKL expression was indeed evident in keratinocytes of the skin and strongly upregulated following UV irradiation. Importantly, in stark contrast to transgenic overexpression of CD40L, RANKL overexpression in keratinocytes abrogated cutaneous contact hypersensitivity responses and concomitantly resulted in a marked increase of Tregs [43]. The receptor for RANKL, RANK, is expressed on LCs and enhanced signaling between RANKL-overexpressing keratinocytes and RANK-expressing LCs increased their survival and rendered LCs more effective in enhancing Treg proliferation [43]. Moreover, RANKL overexpression in keratinocytes could rescue the autoimmunity phenotype caused by CD40L overexpression in K14-RANKL/CD40L double transgenic mice [43]. Taken together, these findings provide a rationale for the long known immunosuppressive effect of ultraviolet exposure: UV irradiation is thought to upregulate RANKL in keratinocytes which in turn activates RANK-expressing LCs through RANKL-RANK interactions. RANKL-activated LCs preferentially trigger expansion of Tregs and thereby suppress immune reactions in the skin and other tissues. Importantly, UV-mediated immunosuppression, as determined by a DTH reaction in the ear, is impaired in mice transplanted with RANKL knock-out skin. Thus, RANKL-RANK might be the missing links to solve a long known conundrum – how sun exposure (sun burns) can be immunosuppressive.

These findings have several important clinical implications: for instance, local induction of the RANKL-RANK system in the skin could be used as a new approach for the treatment of allergies or systemic autoimmunity through increasing Treg numbers while avoiding systemic side effects [43]. The importance of Tregs and the influence of RANKL-RANK signals on their number and regulatory capacity, respectively, has also been previously suggested in an inflammation-induced model of type 1 diabetes. In this mouse model, islet-specific expression of TNF-α can be switched off upon doxycycline administration (Tet-TNF-α); in Tet-TNF-α/CD80 double transgenic mice, which constitutively co-express TNF-α and the costimulatory molecule CD80 on β cells in the islets of Langerhans, progression to diabetes depends on the duration of TNF-α expression [44, 45]. In this model, inflammation activates self-reactive CD8$^+$ T cells to induce autoimmunity and diabetes, respectively, but CD4$^+$25$^+$ Tregs can successfully prevent β cell destruction. These Tregs have been shown to accumulate preferentially in the pancreatic lymph nodes (PLN) and islets and their capacity to prevent diabetes development appears to depend on RANKL signals [45]. Blockade of RANKL-RANK signaling by application of a RANK-Fc resulted in a decreased frequency of CD4$^+$25$^+$ Tregs in the PLN, consequently resulting in intra-islet differentiation of CD8$^+$ T cells into cytotoxic T cells and rapid progression to diabetes [45]. In summary, inflammation may result in the RANKL-RANK-dependent generation and activation of CD4$^+$25$^+$ Treg cells, which then localize to the inflamed tissue and draining lymph nodes for the prevention of tissue destruction and autoimmunity by autoaggressive T cells.

3 T Cells and Bone Loss

Bone remodeling and bone loss are controlled by the RANKL-RANK-OPG axis. Moreover, RANKL is also induced in T cells following antigen receptor engagement. While piecing these findings together, it was intriguing to ask if T cell-derived RANKL could also regulate the development and activation of osteoclasts (i.e., would activated T cells modulate bone turnover via RANKL?). In an in vitro cell culture system of haematopoietic bone marrow precursors, we were indeed able to show that activated CD4$^+$ T cells can induce osteoclastogenesis. Conversely, osteoclastogenesis could be blocked by addition of the physiological decoy receptor of RANKL, and OPG and was not dependent on T-cell-derived cytokines such as IL-1 or TNF-α, which could also upregulate RANKL expression in stromal cells [24]. Activated T cells also affect bone physiology in vivo, as judged by the severe osteoporotic phenotype of Ctla4 knockout mice in which T cells are spontaneously activated. Likewise, transferred Ctla4$^{-/-}$ T cells led to a decrease in bone mineral density in lymphocyte-deficient Rag1$^{-/-}$ mice and continued OPG administration to Ctla4$^{-/-}$ mice diminished their osteoporotic phenotype [24, 49]. These results unequivocally established the pivotal role of systemically activated T cells in resorbing bone through upregulation of RANKL, thereby stressing the importance of T cells as crucial mediators of bone loss in vivo. The results provided a novel paradigm for immune cells as regulators of bone physiology and gave birth to the field of osteoimmunology to account for the interplay between the adaptive immune system and bone metabolism. It also gave a new perspective to certain inflammatory or autoimmune diseases such as rheumatoid arthritis.

4 RANKL-RANK as Key Triggers of Bone Loss in Rheumatoid Arthritis

Rheumatoid arthritis (RA) is a common human autoimmune disease and affects around 1% of people. RA is characterized by chronic inflammation of synovial joints, progressive destruction of cartilage and bone, severe joint pain, and ultimately life-long crippling [50]. Since osteoclasts are found at areas of bone erosion in RA patients [51], it was tempting to speculate that RANKL might be a key mediator of bone erosion in RA patients. Moreover, in an adjuvant-induced arthritis model (AdA), activated, RANKL-expressing T cells specific for the eliciting antigen can transfer the disease [52]. Consequently, we initially analyzed the contribution of RANKL to RA in an AdA model in Lewis rats. The AdA condition in rats mimics many of the clinical and pathological features of human RA (i.e., severe inflammation in bone marrow and soft tissues surrounding joints, accompanied by extensive local bone and cartilage destruction, loss of bone mineral density and crippling) [53]. In addition, T cells present in the inflamed joints and draining lymph nodes produce many pro-inflammatory cytokines [54]. Although inhibition of RANKL through OPG did not influence the severity of inflammation, OPG treatment nonetheless abolished the loss of mineral bone in inflamed joints of arthritic rats in a dose-dependent manner. Bone destruction in untreated arthritic animals correlated with a dramatic increase in osteoclast numbers, which was not observed in OPG-treated rats [24]. As a consequence, OPG-treated arthritic rats exhibited minimal loss of cortical and trabecular bone, whereas untreated AdA animals developed severe bone lesions characterized by partial to complete destruction of cortical and trabecular bone and positively affected erosion of the articular cartilages. In further pre-clinical studies in AdA rats, a single OPG injection was in fact shown to inhibit joint erosions for several days and produced sustained antierosive activity after a short course, but is most effective when initiated early in the disease [55]. These results demonstrated the importance of RANKL in mediating joint destruction and bone loss in AdA arthritis.

An important step in the etiology of arthritis is the alteration of cartilage structures leading to cartilage collapse in the joints. It is not quite clear whether cartilage destruction occurs independently of bone loss or whether damage to the subchondral bone indirectly causes cartilage deterioration [56]. In AdA rats, partial or complete erosion of cartilage in the central and peripheral regions of joint surfaces is observed, which can be preserved by OPG administration. Neither cartilage erosion nor matrix degeneration in the centers of joint surfaces occurred in OPG-treated AdA rats [24]. OPG could protect the cartilage by maintaining the underlying subchondral bone and insulating the overlying cartilage from the inflammatory cell infiltrates in the bone marrow. Since both RANKL and RANK are expressed on chondrocytes [12, 46], and Rankl$^{-/-}$ and Rank$^{-/-}$ mice exhibit significant changes in the columnar alignment of chondrocytes at the growth plate [1, 3], it is possible that the RANKL-RANK pathway plays a direct role in cartilage growth and cartilage homeostasis. These data provided the first evidence that inhibition of RANKL activity by OPG can also prevent cartilage destruction, a critical, irreversible step in the pathogenesis of arthritis. It should be noted that arthritis can also develop in the absence of activated T cells, as shown in the K/B$\times$N serum transfer model of spontaneous autoimmunity [57]. Although RANKL-deficient mice still develop inflammation in the K/B$\times$N serum transfer arthritic model, Rankl$^{-/-}$ mice showed a dramatic reduction in bone erosion – in line with the absolute requirement of RANKL for osteoclastogenesis [58]. However, cartilage damage was still observed in both arthritic Rankl$^{-/-}$ and arthritic control mice, but a trend toward milder cartilage damage in the Rankl$^{-/-}$ mice was noted. Thus, it appears that RANKL is not essential for cartilage destruction but clearly plays an as yet unidentified modulatory role [58].

In every rodent model of arthritis examined such as in TNF-α- or collagen-induced arthritis [59], inhibition of RANKL prevented bone erosion [60]. Of importance is that RANKL expression could be detected in inflammatory cells isolated from the synovial fluid of patients with adult or juvenile RA and patients with osteoarthritis while OPG was not detectable [24]. The correlation between RANKL expression in inflamed joints and arthritis appears to be absolute. In order to precisely define the cells producing RANKL, inflammatory synovial fluids were separated into T and non-T cell populations. Consistent with results obtained in rats, both synovial T and non-T cell populations from RA patients expressed RANKL but not OPG, and the capacity of human T cells expressing RANKL to directly induce osteoclastogenesis from human monocytes has been confirmed [61]. Moreover, RANKL expression is also upregulated in rheumatoid synovial fibroblasts, which in turn can efficiently induce osteoclastogenesis in vitro [62]. These data confirm the findings in rodent adjuvant arthritis, and suggest that RANKL signals from T cells and synoviocytes are the principal mediators of bone destruction in human arthritis. Taken together, RANKL is the trigger of bone loss and crippling in all animal models of arthritis studied so far, making RANKL a prime drug f candidate for therapeutic intervention in different forms of arthritis. Recent phase II clinical trials in humans suggest that inhibition of RANKL in human RA patients have no apparent effects on the inflammation but prevent bone loss at the site of inflammation [63].

These findings also provided a molecular explanation for the observed bone loss in many other humans diseases with chronic activation of the immune system such as adult and childhood leukemia [64], chronic infections such as hepatitis C or HIV [65], autoimmune disorders such as diabetes mellitus and lupus erythematosus [66], allergic diseases such as asthma [67], or lytic bone metastases in multiple cancers such as breast cancer [68]. These bone disorders can all cause irreversible crippling and thereby pose a tremendous burden on the quality of life of a huge number of patients. For example, many patients with lupus require hip replacement surgery and essentially all children that survive leukemia experience severe bone loss and growth retardation. In addition, T cell-derived RANKL also contributes to alveolar bone resorption and tooth loss in an animal model that mimics periodontal disease in humans. This was shown by transplanting human peripheral blood lymphocytes from periodontitis patients into immune-compromised NOD/SCID mice and by challenging these mice with a bacterial strain (*Actinobacillus actinomycetemcomitans*) that can cause periodontitis in humans. In response to stimulation by that microorganism, $CD4^+$ T cells upregulated RANKL and induced osteoclastogenesis and bone destruction, respectively. Most importantly, inhibition of RANKL significantly reduced alveolar bone resorption around the teeth [69]. Further experiments showed that blocking RANKL might also help to prevent periodontitis in diabetic patients which are at high risk of developing periodontitis Nonobese diabetic (NOD) mice – the analog of human type 1 diabetes – were orally infected with

A. actinomycetemcomitans and it turned out that diabetic NOD mice manifested significantly higher alveolar bone loss than non-diabetic control mice. The observed bone loss was correlated with pathogen-specific proliferation and RANKL expression in local $CD4^+$ T cells and could be reduced to baseline levels by RANKL inhibition [70]. Taken together, these findings suggest that specific interference with RANKL signaling pathways might be of great therapeutic value for treating inflammatory bone disorders such as human periodontitis or even bone loss in diabetic patients at high risk.

Since disease pathogenesis correlates with the activation of T cells in many osteopenic disorders; the obvious question then arises of why T cells in our body – of which a certain proportion is activated at any time due to fighting off the universe of foreign antigens to which we are permanently exposed – do not cause extensive bone loss? Likewise, in some chronic T cell and TNF-α-mediated diseases such as ankylosis spondylitis [71], T cell activation does not result in bone loss. One mechanism that counteracts RANKL-mediated bone resorption of activated T cells is the upregulation of interferon-γ (IFN-γ) in certain T cell subsets. IFN-γ blocks RANKL-induced osteoclastogenesis in vitro and IFN-$\gamma R^{-/-}$ mice are more prone to osteoclast formation in a model of endotoxin-induced bone resorption than their wildtype littermates [72]. In line with this study, IFN-γ receptor knockout mice also exhibited enhanced severity in the collagen-induced model of T cell-mediated autoimmune arthritis [73]. Mechanistically, IFN-γ activates the ubiquitin-proteasome pathway in osteoclasts, resulting in TRAF6 degradation and therefore blocks RANK signaling. Thus, it appears that IFN-γ can prevent uncontrolled bone loss during inflammatory T-cell responses. Moreover, T cell-derived IL12 alone and IL12 in synergy with IL18 inhibits osteoclast formation in vitro [74], and IL-4 can abrogate osteoclastogenesis through STAT6-dependent inhibition of NF-κB signaling [75]. Thus, multiple T cell-derived cytokines might be able to interfere with RANK(L) signaling and therefore block osteoclastogenesis and osteoclast functions.

A recent report showed that a certain subset of $CD4^+$ T helper cells, namely Th17, function as osteoclastogenic helper T cells [76]. Derived from naive T cells by a distinct mechanism than Th1 or Th2 cells [77], Th17 cells produce IL-17 and are thus responsible for a variety of autoimmune inflammatory effects [78]. Since IL-17 is also a potent inducer of RANKL expression and can found in the synovial fluid from RA patients [79], Th17 cells seem to be the prime candidate for the osteoclastogenic Th cell subset. Indeed, Th17 cells, but not Th1, Th2, or Treg cells, can stimulate osteoclastogenesis in vitro [76]. This study indicates that Th17 cells act as key mediators of bone destruction in RA patients by different means such as stimulation of local inflammation through IL-17, expression of RANKL on themselves, and induction of RANKL on osteoblasts or synovial fibroblasts, thereby contributing to accelerated bone erosin. The positive effect of Th17 cells on osteoclastogenesis is believed to be balanced by Th1 and Th2 cells mainly through their production of the cytokines IFN-γ and IL-4, respectively [76]. Thus, targeting Th17 might also be a powerful approach to prevent bone destruction associated with T cell activation in RA and other inflammatory bone diseases. Further studies will have to clarify the precise relationship and regulatory crosstalk of Th1, Th2, and Th17 subsets.

5 RANKL Inhibition as a New Therapy to Control Bone Loss in Human Patients

Several years ago, a fully human monoclonal IgG_2 antibody to human RANKL, Denosumab, wasdeveloped and is currently in late-stage clinical trials for postmenopausal osteoporosis, cancer-metastases-induced bone loss, and RA [63, 80, 81]. Significantly, the binding of denosumab to RANKL is selective, and Denosumab does not show any signs of cross-reactivity to TNF-α, TNF-β, CD40L, or TRAIL [82]. Subcutaneous application of Denosumab at 3- or 6-month intervals over a period of 12 months to 412 postmenopausal women with low bone mineral density (BMD) in a randomized, placebo-controlled, dose-ranging phase 2 study resulted in a sustained decrease in bone turnover and a rapid increase in BMD [83]. In another 2-year randomized, double-blind, placebo-controlled study with 332 osteoporotic, postmenopausal women, twice-yearly subcutaneous application of Denosumab significantly increased BMD and decreased bone turnover markers in early and later postmenopausal women [80]. In a similar study in patients with breast cancer ($n = 29$) and multiple myeloma ($n = 25$) with radiologically confirmed bone lesions, a single dose of Denosumab resulted in the rapid and sustained decrease of bone turnover [84]. Lastly, in a multicenter, randomized, double-blind, placebo-controlled, phase II study with 218 patients with RA receiving methotrexate treatment, RANKL inhibition by Denosumab also increased BMD and protected from bone loss at the site of the inflamed joint without affecting inflammation per se [63]. In all cases, Denosumab administration was well tolerated and at least as good or superior to current standard medication. However, considering the various in vivo functions of RANKL-RANK, further clinical trials will be required to substantiate the benefits of RANKL inhibition on suppressing bone destruction.

6 Conclusions

The identification of RANKL, its receptor RANK, and the decoy receptor OPG as the key regulators for osteoclast development and the activation of mature osteoclasts has provided the key molecular framework to understand bone physiology and has opened the doors for the development of highly effective and rational drugs to treat bone loss in millions of patients. The finding that RANKL is produced by activated T cells; that activated T cells, in turn, can directly induce osteoclastogenesis also provided a novel molecular paradigm for bone loss associated with diseases having immune system involvement such as T cell leukemias, autoimmunity, various viral infections, RA, or periodontitis. In addition, RANKL-RANK control development of mammary glands in pregnancy and the formation of lymph nodes and $AIRE^+$ thymic medullary epithelial cells. Moreover, RANKL might be the missing link between sun exposure and Treg mediated immunosuppression. Based on all available data, an important notation is that the inhibition of RANKL function might be the most rational therapy to ameliorate many osteopenic conditions and prevent bone destruction and cartilage damage (e.g., in osteoporosis and arthritis, thereby dramatically enhancing the lives of millions of patients).

References

1. Abu-Amer, Y. (2001). IL-4 abrogates osteoclastogenesis through STAT6-dependent inhibition of NF-kappaB. *J Clin Invest*, 107: 1375–1385.
2. Akiyama, T., Shimo, Y., Yanai, H. et al. (2008). The tumor necrosis factor family receptors RANK and CD40 cooperatively establish the thymic medullary microenvironment and self-tolerance. *Immunity*, 29: 423–437.
3. Alimzhanov, M.B., Kuprash, D.V., Kosco-Vilbois, M.H. et al. (1997). Abnormal development of secondary lymphoid tissues in lymphotoxin beta-deficient mice. *Proc Natl Acad Sci U S A*, 94: 9302–9307.
4. Anderson, G., Lane, P.J., & Jenkinson, E.J. (2007). Generating intrathymic microenvironments to establish T-cell tolerance. *Nat Rev Immunol*, 7: 954–963.
5. Anderson, D.M., Maraskovsky, E., Billingsley, W.L. et al. (1997). A homologue of the TNF receptor and its ligand enhance T-cell growth and dendritic-cell function. *Nature*, 390: 175–179.

6. Anderson, M.S., Venanzi, E.S., Klein, L. et al. (2002). Projection of an immunological self shadow within the thymus by the aire protein. *Science*, 298: 1395–1401.
7. Bachmann, M.F., Wong, B.R., Josien, R. et al. (1999). TRANCE, a tumor necrosis factor family member critical for CD40 ligand-independent T helper cell activation. *J Exp Med*, 189: 1025–1031.
8. Banchereau, J., & Steinman, R.M. (1998). Dendritic cells and the control of immunity. *Nature*, 392: 245–252.
9. Bendele, A., McComb, J., Gould, T. et al. (1999). Animal models of arthritis: relevance to human disease. *Toxicol Pathol*, 27: 134–142.
10. Bendixen, A.C., Shevde, N.K., Dienger, K.M. et al. (2001). IL-4 inhibits osteoclast formation through a direct action on osteoclast precursors via peroxisome proliferator-activated receptor gamma 1. *Proc Natl Acad Sci USA*, 98: 2443–2448; Akiyama, T., Maeda, S., Yamane, S. et al. (2005). Dependence of self-tolerance on TRAF6-directed development of thymic stroma. *Science*, 308: 248–251.
11. Boehm, T., Scheu, S., Pfeffer, K. et al. (2003). Thymic medullary epithelial cell differentiation, thymocyte emigration, and the control of autoimmunity require lympho-epithelial cross talk via LTbetaR. *J Exp Med*, 198: 757–769.
12. Body, J.J., Facon, T., Coleman, R.E. et al. (2006). A study of the biological receptor activator of nuclear factor-kappaB ligand inhibitor, denosumab, in patients with multiple myeloma or bone metastases from breast cancer. *Clin Cancer Res*, 12: 1221–1228.
13. Bone, H.G., Bolognese, M.A., Yuen, C.K. et al. (2008). Effects of denosumab on bone mineral density and bone turnover in postmenopausal women. *J Clin Endocrinol Metab*, 93: 2149–2157.
14. Brandt, J., Haibel, H., Cornely, D. et al. (2000). Successful treatment of active ankylosing spondylitis with the anti-tumor necrosis factor alpha monoclonal antibody infliximab. *Arthritis Rheum*, 43: 1346–1352.
15. Campagnuolo, G., Bolon, B., & Feige, U. (2002). Kinetics of bone protection by recombinant osteoprotegerin therapy in Lewis rats with adjuvant arthritis. *Arthritis Rheum*, 46: 1926–1936; Bolon, B., Campagnuolo, G., & Feige, U. (2002). Duration of bone protection by a single osteoprotegerin injection in rats with adjuvant-induced arthritis. *Cell Mol Life Sci*, 59: 1569–1576.
16. Chin, R.K., Lo, J.C., Kim, O. et al. (2003). Lymphotoxin pathway directs thymic Aire expression. *Nat Immunol*, 4: 1121–1127.
17. Cohen, S.B., Dore, R.K., Lane, N.E. et al. (2008). Denosumab treatment effects on structural damage, bone mineral density, and bone turnover in rheumatoid arthritis: a twelve-month, multicenter, randomized, double-blind, placebo-controlled, phase II clinical trial. *Arthritis Rheum*, 58: 1299–1309.
18. De Togni, P., Goellner, J., Ruddle, N.H. et al. (1994). Abnormal development of peripheral lymphoid organs in mice deficient in lymphotoxin. *Science*, 264: 703–707.
19. Dong, C. (2006). Diversification of T-helper-cell lineages: finding the family root of IL-17-producing cells. *Nat Rev Immunol*, 6: 329–333.
20. Dougall, W.C., & Chaisson, M. (2006). The RANK/RANKL/OPG triad in cancer-induced bone diseases. *Cancer Metastasis Rev*, 25: 541–549.
21. Dougall, W.C., Glaccum, M., Charrier, K. et al. (1999). RANK is essential for osteoclast and lymph node development. *Genes Dev*, 13: 2412–2424.
22. Ebeling, P.R., Erbas, B., Hopper, J.L. et al. (1998). Bone mineral density and bone turnover in asthmatics treated with long-term inhaled or oral glucocorticoids. *J Bone Miner Res*, 13: 1283–1289.
23. Emery, J.G., McDonnell, P., Burke, M.B. et al. (1998). Osteoprotegerin is a receptor for the cytotoxic ligand TRAIL. *J Biol Chem*, 273: 14363–14367.
24. Fata, J.E., Kong, Y.Y., Li, J. et al. (2000). The osteoclast differentiation factor osteoprotegerin-ligand is essential for mammary gland development. *Cell*, 103: 41–50.

25. Feldmann, M., Brennan, F.M., & Maini, R.N. (1996). Rheumatoid arthritis. *Cell*, 85: 307–310.
26. Feldmann, M., Brennan, F.M., & Maini, R.N. (1996). Role of cytokines in rheumatoid arthritis. *Annu Rev Immunol*, 14: 397–440.
27. Futterer, A., Mink, K., Luz, A. et al. (1998). The lymphotoxin beta receptor controls organogenesis and affinity maturation in peripheral lymphoid tissues. *Immunity*, 9: 59–70.
28. Fu, Y.X., Chaplin, D.D. (1999). Development and maturation of secondary lymphoid tissues. *Annu Rev Immunol*, 17: 399–433.
29. Green, E.A., Choi, Y., & Flavell, R.A. (2002). Pancreatic lymph node-derived CD4(+)CD25(+) Treg cells: highly potent regulators of diabetes that require TRANCE-RANK signals. *Immunity*, 16: 183–191.
30. Green, E.A., & Flavell, R.A. (2000). The temporal importance of TNFalpha expression in the development of diabetes. *Immunity*, 12: 459–469.
31. Harrington, L.E., Hatton, R.D., Mangan, P.R. et al. (2005). Interleukin 17-producing CD4+ effector T cells develop via a lineage distinct from the T helper type 1 and 2 lineages. *Nat Immunol*, 6: 1123–1132.
32. Hikosaka, Y., Nitta, T., Ohigashi, I. et al. (2008). The cytokine RANKL produced by positively selected thymocytes fosters medullary thymic epithelial cells that express autoimmune regulator. *Immunity*, 29: 438–450.
33. Hofbauer, L.C., Khosla, S., Dunstan, C.R. et al. (1999). Estrogen stimulates gene expression and protein production of osteoprotegerin in human osteoblastic cells. *Endocrinology*, 140: 4367–4370.
34. Horwood, N.J., Elliott, J., Martin, T.J. et al. (2001). IL-12 alone and in synergy with IL-18 inhibits osteoclast formation in vitro. *J Immunol*, 166: 4915–4921.
35. Hsu, H., Lacey, D.L., Dunstan, C.R. et al. (1999). Tumor necrosis factor receptor family member RANK mediates osteoclast differentiation and activation induced by osteoprotegerin ligand. *Proc Natl Acad Sci U S A*, 96: 3540–3545.
36. Ingulli, E., Mondino, A., Khoruts, A. et al. (1997). In vivo detection of dendritic cell antigen presentation to CD4(+) T cells. *J Exp Med*, 185: 2133–2141
37. Irla, M., Hugues, S., Gill, J. et al. (2008). Autoantigen-specific interactions with CD4+ thymocytes control mature medullary thymic epithelial cell cellularity. *Immunity*, 29: 451–463.
38. Josien, R., Li, H.L., Ingulli, E. et al. (2000). TRANCE, a tumor necrosis factor family member, enhances the longevity and adjuvant properties of dendritic cells in vivo. *J Exp Med*, 191: 495–502.
39. Josien, R., Wong, B.R., Li, H.L. et al. (1999). TRANCE, a TNF family member, is differentially expressed on T cell subsets and induces cytokine production in dendritic cells. *J Immunol*, 162: 2562–2568.
40. Kanematsu, M., Sato, T., Takai, H., et al. (2000). Prostaglandin E2 induces expression of receptor activator of nuclear factor-kappa B ligand/osteoprotegrin ligand on pre-B cells: implications for accelerated osteoclastogenesis in estrogen deficiency. *J Bone Miner Res*, 15: 1321–1329.
41. Keffer, J., Probert, L., Cazlaris, H. et al (1991). Transgenic mice expressing human tumour necrosis factor: a predictive genetic model of arthritis. *Embo J*, 10: 4025–4031,
42. Kim, D., Mebius, R.E., MacMicking, J.D. et al. (2000). Regulation of peripheral lymph node genesis by the tumor necrosis factor family member TRANCE. *J Exp Med*, 192: 1467–1478.
43. Kong, Y.Y., Feige, U., Sarosi, I. et al. (1999). Activated T cells regulate bone loss and joint destruction in adjuvant arthritis through osteoprotegerin ligand. *Nature*, 42: 304–309.
44. Kong, Y.Y., Yoshida, H., Sarosi, I. et al. (1999). OPGL is a key regulator of osteoclastogenesis, lymphocyte development and lymph-node organogenesis. *Nature*, 397: 315–323.

45. Koni, P.A., Sacca, R., Lawton, P. et al. (1997). Distinct roles in lymphoid organogenesis for lymphotoxins alpha and beta revealed in lymphotoxin beta-deficient mice. *Immunity*, 6: 491–500.
46. Korganow, A.S., Ji, H., Mangialaio, S. et al. (1999). From systemic T cell self-reactivity to organ-specific autoimmune disease via immunoglobulins. *Immunity*, 10: 451–461.
47. Kotake, S., Udagawa, N., Hakoda, M. et al. (2001). Activated human T cells directly induce osteoclastogenesis from human monocytes: possible role of T cells in bone destruction in rheumatoid arthritis patients. *Arthritis Rheum*, 44: 1003–1012.
48. Kotake, S., Udagawa, N., Takahashi, N. et al. (1999). IL-17 in synovial fluids from patients with rheumatoid arthritis is a potent stimulator of osteoclastogenesis. *J Clin Invest*, 103: 1345–1352.
49. Kouskoff, V., Korganow, A.S., Duchatelle, V., &et al. (1996). Organ-specific disease provoked by systemic autoimmunity. *Cell*, 87: 811–822.
50. Kyewski, B., & Klein, L. (2006). A central role for central tolerance. *Annu Rev Immunol*, 24: 571–606.
51. Lacey, D.L., Timms, E., Tan, H.L. et al. (1998). Osteoprotegerin ligand is a cytokine that regulates osteoclast differentiation and activation. *Cell*, 93: 165–176.
52. Li, J., Sarosi, I., Yan, X.Q. et al. (2000). RANK is the intrinsic hematopoietic cell surface receptor that controls osteoclastogenesis and regulation of bone mass and calcium metabolism. *Proc Natl Acad Sci U S A*, 97: 1566–1571.
53. Lipton, A. (2006). Future treatment of bone metastases. *Clin Cancer Res*, 12: 6305s–6308s.
54. Lipton, A., Steger, G.G., Figueroa, J. et al. (2007). Randomized active-controlled phase II study of denosumab efficacy and safety in patients with breast cancer-related bone metastases. *J Clin Oncol*, 25: 4431–4437.
55. Loser, K., Mehling, A., Loeser, S. et al. (2006). Epidermal RANKL controls regulatory T-cell numbers via activation of dendritic cells. *Nat Med*, 12: 1372–1379.
56. Mahamed, D.A., Marleau, A., Alnaeeli, M. et al. (2005). G(-) anaerobes-reactive CD4+ T-cells trigger RANKL-mediated enhanced alveolar bone loss in diabetic NOD mice. *Diabetes*, 54: 1477–1486.
57. Manoury-Schwartz, B., Chiocchia, G., Bessis, N. et al. (1997). High susceptibility to collagen-induced arthritis in mice lacking IFN-gamma receptors. *J Immunol*, 158: 5501–5506.
58. Matsumoto, M., Mariathasan, S., Nahm, M.H. et al. (1996). Role of lymphotoxin and the type I TNF receptor in the formation of germinal centers. *Science*, 271: 1289–1291.
59. Mebius, R.E. (2003). Organogenesis of lymphoid tissues. *Nat Rev Immunol*, 3: 292–303.
60. Mebius, R.E., Rennert, P., & Weissman, I.L. (1997). Developing lymph nodes collect CD4+CD3- LTbeta+ cells that can differentiate to APC, NK cells, and follicular cells but not T or B cells. *Immunity*, 7: 493–504.
61. Mehling, A., Loser, K., Varga, G. et al. (2001). Overexpression of CD40 ligand in murine epidermis results in chronic skin inflammation and systemic autoimmunity. *J Exp Med*, 194: 615–628.
62. McClung, M.R., Lewiecki, E.M., Cohen, S.B. et al. (2006). Denosumab in postmenopausal women with low bone mineral density. *N Engl J Med*, 354: 821–831.
63. Mori, H., Kitazawa, R., Mizuki, S. et al. (2002). RANK ligand, RANK, and OPG expression in type II collagen-induced arthritis mouse. *Histochem Cell Biol*, 117: 283–292.
64. Muller-Ladner, U., Gay, R.E., & Gay, S. (1998). Molecular biology of cartilage and bone destruction. *Curr Opin Rheumatol*, 10: 212–219.
65. Nakashima, T., Wada, T., & Penninger, J.M. (2003). RANKL and RANK as novel therapeutic targets for arthritis. *Curr Opin Rheumatol*, 15: 280–287.
66. Oliveri, M.B., Mautalen, C.A., Rodriguez Fuchs, C.A. et al. (1991). Vertebral compression fractures at the onset of acute lymphoblastic leukemia in a child. *Henry Ford Hosp Med J*, 39: 45–48.

67. Oxenius, A., Campbell, K.A., Maliszewski, C.R. et al. (1996). CD40-CD40 ligand interactions are critical in T-B cooperation but not for other anti-viral CD4+ T cell functions. *J Exp Med*, 183: 2209–2218.
68. Oliveri, MB., Mautalen, C.A., Rodriguez Fuchs, C.A. et al. (1991). Vertebral compression fractures at the onset of acute lymphoblastic leukemia in a child. *Henry Ford Hosp Med J*, 39: 45–48.
69. Oxenius, A., Campbell, K.A., Maliszewski, C.R. et al. (1996). CD40-CD40 ligand interactions are critical in T-B cooperation but not for other anti-viral CD4+ T cell functions. *J Exp Med*, 183: 2209–2218.
70. Panayi, G.S., Lanchbury, J.S., & Kingsley, G.H. (1992). The importance of the T cell in initiating and maintaining the chronic synovitis of rheumatoid arthritis. *Arthritis Rheum*, 35: 729–735.
71. Park, H., Li, Z., Yang, X.O. et al. (2005). A distinct lineage of CD4 T cells regulates tissue inflammation by producing interleukin 17. *Nat Immunol*, 6: 1133–1141.
72. Pettit, A.R., Ji, H., von Stechow, D. et al. (2001). TRANCE/RANKL knockout mice are protected from bone erosion in a serum transfer model of arthritis. *Am J Pathol*, 159: 1689–1699.
73. Piepkorn, B., Kann, P., Forst, T. et al. (1997). Bone mineral density and bone metabolism in diabetes mellitus. *Horm Metab Res*, 29: 584–591.
74. Quezada, S.A., Jarvinen, L.Z., Lind, E.F. et al. (2004). CD40/CD154 interactions at the interface of tolerance and immunity. *Annu Rev Immunol*, 22: 307–328.
75. Redlich, K., Hayer, S., Maier, A. et al. (2002). Tumor necrosis factor alpha-mediated joint destruction is inhibited by targeting osteoclasts with osteoprotegerin. *Arthritis Rheum*, 46: 785–792.
76. Rennert, P.D., Browning, J.L., & Hochman, P.S. (1997). Selective disruption of lymphotoxin ligands reveals a novel set of mucosal lymph nodes and unique effects on lymph node cellular organization. *Int Immunol*, 9: 1627–1639.
77. Rennert, P.D., James, D., Mackay, F. et al. (1998). Lymph node genesis is induced by signaling through the lymphotoxin beta receptor. *Immunity*, 9: 71–79.
78. Romas, E., Sims, N.A., Hards, D.K. et al. (2002). Osteoprotegerin reduces osteoclast numbers and prevents bone erosion in collagen-induced arthritis. *Am J Pathol*, 161: 1419–1427.
79. Rossi, S.W., Jenkinson, W.E., Anderson, G. et al. (2006). Clonal analysis reveals a common progenitor for thymic cortical and medullary epithelium. *Nature*, 441: 988–991.
80. Rossi, S.W., Kim, M.Y., Leibbrandt, A. et al. (2007). RANK signals from CD4(+)3(-) inducer cells regulate development of Aire-expressing epithelial cells in the thymic medulla. *J Exp Med*, 204: 1267–1272.
81. Roy, M., Waldschmidt, T., Aruffo, A. et al. (1993). The regulation of the expression of gp39, the CD40 ligand, on normal and cloned CD4+ T cells. *J Immunol*, 151: 2497–2510.
82. Sakaguchi, S. (2005). Naturally arising Foxp3-expressing CD25+CD4+ regulatory T cells in immunological tolerance to self and non-self. *Nat Immunol*, 6: 345–352.
83. Sato, K., Suematsu, A., Okamoto, K. et al. (2006). Th17 functions as an osteoclastogenic helper T cell subset that links T cell activation and bone destruction. *J Exp Med*, 203: 2673–2682.
84. Seitz, M., & Hunstein, W. (1985). Enhanced prostanoid release from monocytes of patients with rheumatoid arthritis and active systemic lupus erythematosus. *Ann Rheum Dis*, 44: 438–445.

Eph and Ephrin Interactions in Bone

Koichi Matsuo

Abstract Bone cells communicate with each other using various cell surface molecules. Membrane-bound ephrin ligands and Eph tyrosine kinase receptors have been characterized in diverse biological processes, including angiogenesis and neuronal development. Several ephrins and Ephs are expressed in osteoclasts and osteoblasts and regulate bone mineral metabolism through bidirectional signaling into not only receptor-expressing cells but also into ligand-expressing cells. We propose that interaction between ephrinB2-expressing osteoclasts and EphB4-expressing osteoblasts facilitates the transition from bone resorption to bone formation during bone remodeling. Other groups have reported the regulation of ephrinB2 by PTH or PTHrP and the possible involvement of EphB4 in osteoarthritis. It is likely that various ephrins and Ephs mediate interaction among bone cells.

Keywords Eph · Ephrin · Osteoclast · Bone remodeling

1 Introduction

Bone-resorbing osteoclasts and bone-forming osteoblasts have two different lineages: hematopoietic and mesenchymal stem cells, respectively. Osteoclasts are polarized, multinucleated macrophages that specialize in bone resorption and that share precursor cells with mononuclear macrophages that engulf bacteria, viruses, or parasites [19]. Osteoclasts become multinucleated through cell–cell fusion. Unlike Osteoclasts, bone-forming osteoblasts are mononuclear cells derived from mesenchymal progenitors that also differentiate into myoblasts, chondrocytes, adipocytes, and fibroblasts. The overall balance between bone resorption and bone formation in the body, conceptualized as "coupling," is essential for maintaining

K. Matsuo (✉)
Collaborative Research Resources, School of Medicine, Keio University, 160-8582 Tokyo, Japan
e-mail: matsuo@sc.itc.keio.ac.jp

Y. Choi (ed.), *Osteoimmunology*, Advances in Experimental Medicine and Biology 658, DOI 10.1007/978-1-4419-1050-9_10,

bone mass. Such balance is thought to be the sum of cellular activities, which are regulated both locally and systemically by hormones and the nervous system. Bone remodeling occurs sequentially: osteoclastic bone resorption is followed by an equivalent amount of bone formation, which refills the resorption lacunae or pits [8].

Differentiation of osteoclasts is known to depend on osteoblasts or stromal cells. In fact, hematopoietic precursors such as non-adherent bone marrow cells or spleen cells can differentiate into osteoclasts when they are co-cultured with osteoblasts isolated from fetal or neonatal mouse calvaria [25]. This osteoclast-forming activity is mainly explained by two molecules expressed on osteoblasts: macrophage colony stimulating factor (M-CSF) and RANKL. In fact, soluble forms of M-CSF and RANKL in place of osteoblasts allow in vitro osteoclastogenesis [6, 17, 27]. In addition to M-CSF and RANKL, osteoblasts also produce so-called "costimulatory molecules" to activate OSCAR and other receptors on osteoclast precursors [12, 13].

Cellular interaction between the osteoclast and osteoblast lineages is not limited to the induction of ostoeclastogenesis by osteoblasts. After osteoclasts differentiate, mature ones likely induce the differentiation of osteoblast progenitors (see below). Even mature osteoclasts and mature osteoblasts can be found next to each other [18]. Various soluble and membrane-bound factors produced by osteoclasts can act on osteoblasts and vice versa. We are working on the roles of ephrin ligands and Eph receptors in bone cell communication. Accumulating data including our own suggest that various ephrin ligands and Eph receptors mediate communication between and within cells in osteoclast and osteoblast lineages.

2 Ephrin-Eph Family

Eph receptors are tyrosine kinase receptors. The name Eph derives from this family's first member, which was cloned from an erythropoietin-producing hepatocellular carcinoma cell line [5]. Eph receptors are classified into classes A and B. In mammals, there are nine EphA receptors (EphA1–EphA9) and five EphB receptors (EphB1 to EphB6, no EphB5). The name ephrin is rooted in the ancient Greek έ φ ο ρ ο ς (ephoros), meaning overseer or controller, and it is an acronym for Eph family receptor interacting protein [5]. The ephrinA ligands (ephrinA1–A5) are glycosylphosphatidylinositol (GPI)-anchored and interact with EphA receptors while ephrinB ligands (ephrinB1–B3), which have transmembrane domains, interact with EphB receptors (Fig. 1). These ephrin ligands and Eph receptors have been analyzed in the nervous system, vasculature, and digestive organs, among others [22]. Both ephrinB/EphB and ephrinA/EphA systems generate bidirectional signals that affect ligand-expressing cells in addition to receptor-expressing cells. These signals regulate cell–cell boundaries, cell adhesion and repulsion, cell shape, and cell migration. Recently, evidence has accumulated that ephrin-Eph molecules have significant roles in bone biology.

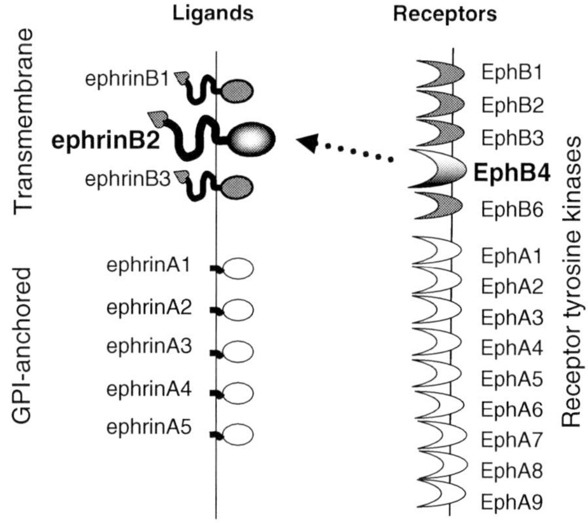

Fig. 1 The ephrin and Eph family members. Note that ephrinB2 is known as a preferred ligand for EphB4 receptor. Both the ephrinB/EphB and ephrinA/EphA systems signal bidirectionally, affecting both the ligand-expressing and receptor-expressing cells

3 EphrinB2-EphB4 in Osteoclasts and Osteoblasts

3.1 Coupling Between Bone Resorption and Bone Formation

EphrinB2 is the preferred ligand for the EphB4 receptor. We found that differentiating osteoclasts induce ephrinB2 expression and that the reverse signaling through ephrinB2 suppresses osteoclast differentiation, forming a negative feedback loop [28]. Moreover, osteoblasts express receptor EphB4, and the forward signaling through EphB4 into osteoblasts enhances osteoblast differentiation. The in vitro cell culture experiments and in vivo transgenic mouse experiments consistently support the view that ephrinB2-EphB4 interaction simultaneously suppresses and enhances osteoclastogenesis and osteoblastogenesis, respectively, through bidirectional signaling (Fig. 2). In fact, transgenic mice expressing EphB4 under the control of the type I collagen promoter showed not only increased bone formation but also decreased bone resorption based on dual-energy X-ray absorptiometry (DXA), microcomputed tomography (μCT), bone histomorphometry, quantification of serum osteocalcin, and urinary deoxypyridinoline (DPD)-crosslinks. It is possible that ephrinB2-EphB4 interaction facilitates the transition from bone resorption to bone formation [21]. In this respect, ephrinB2-EphB4 interaction can be considered a coupling factor.

It is important to mention that ephrinB2-EphB4 interaction also occurs between osteoblasts, since osteoblasts can express both ephrinB2 and EphB4 under certain conditions. Intermittent administration of parathyroid hormone (PTH) produces anabolic effects in experimental animals, and PTH is a clinically promising anabolic factor. Martin and colleagues found that PTH induces ephrinB2

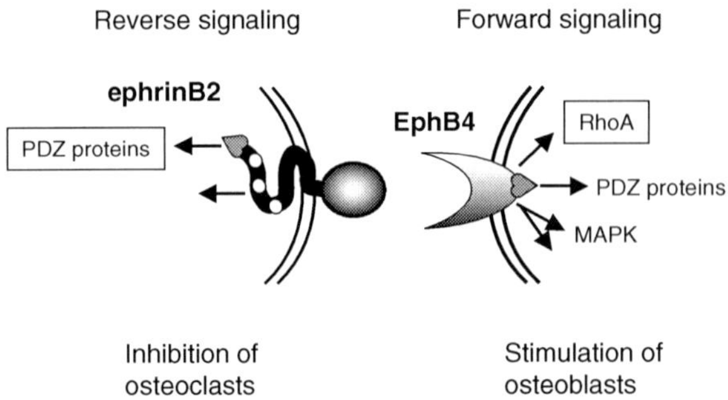

Fig. 2 Interaction between ephrinB2 on osteoclasts and EphB4 on osteoblasts. The reverse signaling through ephrinB2 can occur not only by interacting with PDZ domain proteins at the C-terminal cytoplasmic end, but also by phosphorylating tyrosine residues (*white circles*). Bidirectional signaling suppresses osteoclast differentiation and enhances osteoblast differentiation. Interaction with unknown PDZ domain proteins appears to be critical for the suppression [28]

expression in the osteoblastic cell line UMR and primary clavarial osteoblasts [1]. Since these osteoblasts express EphB4 constitutively, induction of ephrinB2 enhances osteoblast-osteoblast interaction and thereby stimulates bone formation.

3.2 Osteoarthritis and Cancer Induced Bone Disease

Interaction between ephrinB2 and EphB4 is implicated in abnormal subchondral bone metabolism in osteoarthritis (OA). Degradation of articular cartilage is a characteristic of OA; however, changes in the metabolism of subchondral bone that occur at an early stage of OA, appear to be a basis for the degradation and loss of cartilage [23]. EphB4 receptor expression is high in a subclass of human OA subchondral bone osteoblasts that do not produce high levels of prostaglandin E2 (PGE2). Enhanced EphB4 signaling in these osteoblasts inhibits bone resorption directly or by reducing osteoblastic expression of IL-1β, IL-6, matrix metalloproteinase 1 (MMP-1), MMP-9, MMP-13, and RANKL [16]. EphB4 receptor activation by ephrin B2 in OA subchondral bone may also enhance osteoblast differentiation directly.

EphB4 is also implicated in cancer-induced bone disease. For example, myeloma cells downregulate EphB4 expression in osteoblasts [4]. Since EphB4 forward signaling enhances bone formation, reduction in EphB4 may account for impaired bone formation in myeloma bone disease. Therefore, it is likely that metastatic cancer may modulate bone remodeling through ephrinB-EphB4 signaling.

3.3 Monocytes

Although we did not detect EphB receptors in osteoclasts, circulating monocytes express EphB receptors and vascular endothelial cells express the ephrinB2 ligand on the luminal surface. Adhesion and trafficking experiments suggest that interaction between EphB receptors and ephrinB controls monocyte adhesion and transmigration through the vascular endothelium [24]. Interestingly, stretch-induced ephrinB2 expression in endothelial cells limits smooth muscle cell migration and controls monocyte extravasation [14]. Since osteoclast differentiation appears to be regulated at the point of the mobilization from blood vessel to bone surface [10], it will be interesting to analyze the role of ephinB-EphB interaction in mobilization.

3.4 Downstream Signaling of EphrinB2 and EphB4

The intracellular portion of the EphB4 receptor contains the tyrosine kinase domain, the sterile alpha motif (SAM) domain, and the PDZ (named after PSD95, Dlg, and Zo-1) domain-binding site. These domains mediate forward signaling, which regulates various downstream molecules including small GTPase, MAP kinases, c-src, and PI3K-Akt. We observed that stimulation of mouse calvarial osteoblasts with ephrinB2-Fc reduced the amount of active (GTP-bound) RhoA during the 3–5 days after the addition of β-glycerophosphate and ascorbic acid to culture medium to induce osteoblastic differentiation (Fig. 3A). This reduction of active RhoA may be the cause of enhanced osteoblast differentiation, because a constitutively active form

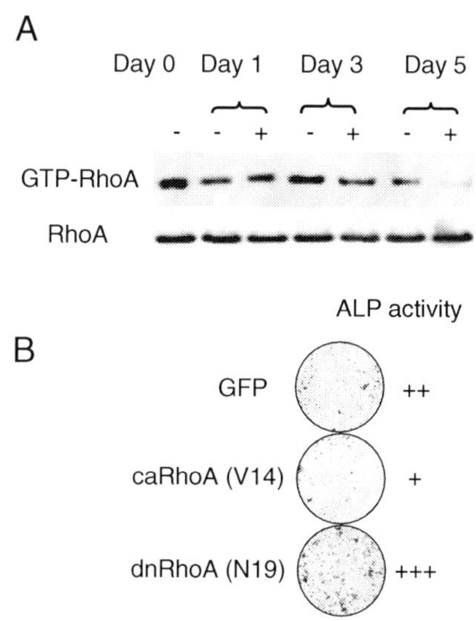

Fig. 3 Analysis of RhoA as a downstream effector of EphB4 forward signaling in osteoblasts. (**A**) Western blot analysis of RhoA activity in calvarial osteoblasts stimulated with Fc (–, negative control) or ephrinB2-Fc (+) for the indicated days. (**B**) Effect of RhoA activity on osteoblast differentiation. Retroviral gene transfer was performed to introduce green fluorescent protein (GFP, negative control), a constitutively active form (V14) of RhoA (caRhoA), or a dominant negative form (N19) of RhoA (dnRhoA). Alkaline phosphatase (ALP) staining at day 6 of calvarial osteoblasts. ALP activity in cell lysates: low (+), intermediate (++), and high (+++)

of RhoA (V14) suppressed osteoblast differentiation when introduced via retroviral gene transfer. Conversely, a dominant negative form of RhoA (N19) enhanced osteoblast differentiation (Fig. 3B). These are consistent with the notion that higher RhoA activity inhibits the differentiation of mouse osteoblasts [7]. Curiously, human osteoblasts behave differently from mouse osteoblasts. In human mesenchymal cells, RhoA enhances rather than suppresses differentiation into the osteoblast lineage [20]. The reason for the human-mouse difference regarding the role of RhoA in osteoblastogenesis is unclear. In mouse cells, we observed that stimulation of EphB4 forward signaling increased ERK1/2 activity on day 6 of osteoblast differentiation (Fig. 4). In human HUVEC cells, the forward signaling of EphB receptors other than EphB4 suppresses ERK1/2 induced by vascular endothelial cell growth factor (VEGF) or angiopoietin 1 [11]. Whether this discrepancy is due to differences in species, cell types, EphB subtypes, or co-existing signaling is unclear at the moment.

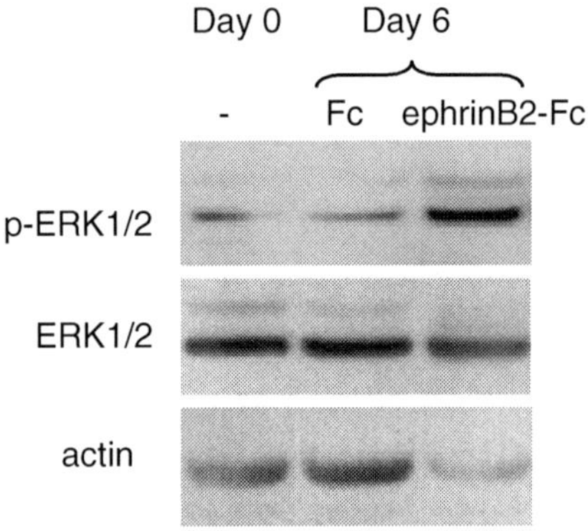

Fig. 4 Analysis of ERK1/2 as downstream effector of EphB4 forward signaling in osteoblasts. Western blotting analysis of ERK phosphorylation (p-ERK1/2) in primary calvarial osteoblasts cultured with Fc or ephrinB2-Fc for 6 days. ERK1/2 denotes total ERK1/2, and actin serves as a loading control

The intracellular portion of ephrinB2 consists of about 80 amino acids (the C-terminal 33 amino acids are highly conserved among ephrinB ligands in mouse and human) and contains well-conserved phosphorylatable tyrosines and a PDZ-domain binding site at the C-terminal end. Reverse signaling through ephrin ligands into ligand-expressing cells decreases c-Fos activity, presumably through the PDZ domain binding site (Fig. 5) [28]. How this divergent downstream signaling is wired to regulate osteoblast and osteoclast differentiation and activation should be analyzed in the future.

4 Roles of Other Ephrin and Eph Family Members in Bone Cells

Mutations in ephrinB1 in humans cause craniofrontonasal syndrome (CFNS), an X-linked disease that affects female patients more severely than males [26]. Mice lacking ephrinB1 are embryonic lethal but heterozygous mutants (ephrinB1+/–)

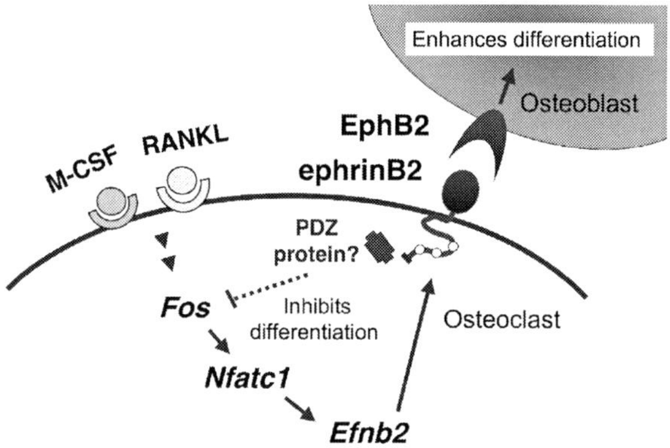

Fig. 5 Schematic presentation of a negative feedback loop involving ephrinB2 reverse signaling in osteoclasts. The c-Fos-NFATc1 transcriptional cascade induces ephrinB2, and ephrinB2-EphB4 interaction suppresses c-Fos transcription through unknown PDZ domain proteins

show phenotypes reminiscent of CFNS [2]. EphrinB1 interacts with connexin43 and affects gap junction communication [3]. Recently, we and others have begun to determine the roles of class A ephrins and Ephs in bone. We have shown that ephrinA ligands and EphA receptors regulate the initiation phase of bone remodeling, especially during osteoclast differentiation. EphrinA2 expression was rapidly induced by RANKL in osteoclast precursors, while ephrinB2 expression increased gradually toward the end of osteoclast differentiation [9]. In addition to ephrinA2, EphA2, a receptor for ephrinA2, was also expressed in both osteoclast precursors and osteoblasts. Therefore, osteoclast precursors can interact with not only osteoblasts but also other osteoclast-lineage cells through ephrinA2-EphA2. EphA4 is implicated in the regulation of ossification [15].

5 Future Directions

First, analyses of interaction between bone cells through ephrin-Eph have focused mainly on osteoclasts and osteoblasts. However, these cells also interact with chondrocytes, osteocytes, and even non-bone cells such as neurons, glia, or endothelial cells. So far, little is known about communication between bone cells and non-bone cells through ephrins and Ephs. This should be explored in the future. Second, intracellular signaling downstream of ephrins and Ephs in bone cells needs to be analyzed, especially in relation to other major signals in these cells, including RANK signaling in osteoclasts and Wnt signaling in osteoblasts. Third, we observed that class A ephrins and Ephs are also involved in the regulation of bone remodeling. Further experiments both in vitro and in vivo are necessary to establish functions of class A ephrin-Eph interactions in bone.

References

1. Allan, E. H., Hausler, K. D., Wei, T., Gooi, J. H., Quinn, J. M., Crimeen-Irwin, B., Pompolo, S. et al. (2008). EphrinB2 regulation by PTH and PTHrP revealed by molecular profiling in differentiating osteoblasts. *J Bone Miner Res*, 23, 1170–1181.
2. Compagni, A., Logan, M., Klein, R., & Adams, R. H. (2003). Control of skeletal patterning by ephrinB1-EphB interactions. *Dev Cell*, 5, 217–230.
3. Davy, A., Bush, J. O., & Soriano, P. (2006). Inhibition of gap junction communication at ectopic Eph/ephrin boundaries underlies craniofrontonasal syndrome. *PLoS Biol*, 4, e315.
4. Edwards, C. M., & Mundy, G. R. (2008). Eph receptors and ephrin signaling pathways: a role in bone homeostasis. *Int J Med Sci*, 5, 263–272.
5. Eph Nomenclature Committee (1997). Unified nomenclature for Eph family receptors and their ligands, the ephrins. Eph Nomenclature Committee. *Cell*, 90, 403–404.
6. Fuller, K., Wong, B., Fox, S., Choi, Y., & Chambers, T. J. (1998). TRANCE is necessary and sufficient for osteoblast-mediated activation of bone resorption in osteoclasts. *J Exp Med*, 188, 997–1001.
7. Harmey, D., Stenbeck, G., Nobes, C. D., Lax, A. J., & Grigoriadis, A. E. (2004). Regulation of osteoblast differentiation by Pasteurella multocida toxin (PMT): a role for Rho GTPase in bone formation. *J Bone Miner Res*, 19, 661–670.
8. Hattner, R., Epker, B. N., & Frost, H. M. (1965). Suggested sequential mode of control of changes in cell behaviour in adult bone remodelling. *Nature*, 206, 489–490.
9. Irie, N., Takada, Y., Watanabe, Y., Matsuzaki, Y., Naruse, C., Asano, M., Iwakura, Y. Sudha, T., & Matsuo, K. (2009). Bidirectional signaling through ephrinA2-EphA2 enhances osteoclastogenesis and suppresses osteoblastogenesis. *J Biol Chem*, 284, 14637–14644.
10. Ishii, M., Egen, J. G., Klauschen, F., Meier-Schellersheim, M., Saeki, Y., Vacher, J., Proia, R.L., & Germain, R. N. (2009). Sphingosine-1-phosphate mobilizes osteoclast precursors and regulates bone homeostasis. *Nature*, 458, 524–528.
11. Kim, I., Ryu, Y. S., Kwak, H. J., Ahn, S. Y., Oh, J. L., Yancopoulos, G. D. et al. (2002). EphB ligand, ephrinB2, suppresses the VEGF- and angiopoietin 1-induced Ras/mitogen-activated protein kinase pathway in venous endothelial cells. *Faseb J*, 16, 1126–1128.
12. Kim, Y., Sato, K., Asagiri, M., Morita, I., Soma, K., & Takayanagi, H. (2005). Contribution of nuclear factor of activated T cells c1 to the transcriptional control of immunoreceptor osteoclast-associated receptor but not triggering receptor expressed by myeloid cells-2 during osteoclastogenesis. *J Biol Chem*, 280, 32905–32913.
13. Koga, T., Inui, M., Inoue, K., Kim, S., Suematsu, A., Kobayashi, E. et al. (2004). Costimulatory signals mediated by the ITAM motif cooperate with RANKL for bone homeostasis. *Nature*, 428, 758–763.
14. Korff, T., Braun, J., Pfaff, D., Augustin, H. G., & Hecker, M. (2008). Role of ephrinB2 expression in endothelial cells during arteriogenesis: impact on smooth muscle cell migration and monocyte recruitment. *Blood*, 112, 73–81.
15. Kuroda, C., Kubota, S., Kawata, K., Aoyama, E., Sumiyoshi, K., Oka, M. et al. (2008). Distribution, gene expression, and functional role of EphA4 during ossification. *Biochem Biophys Res Commun*, 374, 22–27.
16. Kwan Tat, S., Pelletier, J. P., Amiable, N., Boileau, C., Lajeunesse, D., Duval, N., & Martel-Pelletier, J. (2008). Activation of the receptor EphB4 by its specific ligand ephrin B2 in human osteoarthritic subchondral bone osteoblasts. *Arthritis Rheum*, 58, 3820–3830.
17. Lacey, D. L., Timms, E., Tan, H. L., Kelley, M. J., Dunstan, C. R., Burgess, T. et al. (1998). Osteoprotegerin ligand is a cytokine that regulates osteoclast differentiation and activation. *Cell*, 93, 165–176.
18. Matsuo, K., & Irie, N. (2008). Osteoclast-osteoblast communication. *Arch Biochem Biophys*, 473, 201–209.
19. Matsuo, K., & Ray, N. (2004). Osteoclasts, mononuclear phagocytes, and c-Fos: new insight into osteoimmunology. *Keio J Med*, 53, 78–84.

20. McBeath, R., Pirone, D. M., Nelson, C. M., Bhadriraju, K., & Chen, C. S. (2004). Cell shape, cytoskeletal tension, and RhoA regulate stem cell lineage commitment. *Dev Cell*, 6, 483–495.
21. Mundy, G. R., & Elefteriou, F. (2006). Boning up on ephrin signaling. *Cell*, 126, 441–443.
22. Pasquale, E. B. (2008). Eph-ephrin bidirectional signaling in physiology and disease. *Cell*, 133, 38–52.
23. Pelletier, J. P., Boileau, C., Brunet, J., Boily, M., Lajeunesse, D., Reboul, P., Laufer, S., & Martel-Pelletier, J. (2004). The inhibition of subchondral bone resorption in the early phase of experimental dog osteoarthritis by licofelone is associated with a reduction in the synthesis of MMP-13 and cathepsin K. *Bone*, 34, 527–538.
24. Pfaff, D., Heroult, M., Riedel, M., Reiss, Y., Kirmse, R., Ludwig, T., Korff, T., Hecker, M., & Augustin, H. G. (2008). Involvement of endothelial ephrin-B2 in adhesion and transmigration of EphB-receptor-expressing monocytes. *J Cell Sci*, 121, 3842–3850.
25. Takahashi, N., Akatsu, T., Udagawa, N., Sasaki, T., Yamaguchi, A., Moseley, J. M., Martin, T. J., & Suda, T. (1988). Osteoblastic cells are involved in osteoclast formation. *Endocrinology*, 123, 2600–2602.
26. Twigg, S. R., Kan, R., Babbs, C., Bochukova, E. G., Robertson, S. P., Wall, S. A., Morriss-Kay, G. M., & Wilkie, A. O. (2004). Mutations of ephrin-B1 (EFNB1), a marker of tissue boundary formation, cause craniofrontonasal syndrome. *Proc Natl Acad Sci U S A*, 101, 8652–8657.
27. Yasuda, H., Shima, N., Nakagawa, N., Yamaguchi, K., Kinosaki, M., Mochizuki, S. et al. (1998). Osteoclast differentiation factor is a ligand for osteoprotegerin/osteoclastogenesis-inhibitory factor and is identical to TRANCE/RANKL. *Proc Natl Acad Sci U S A*, 95, 3597–3602.
28. Zhao, C., Irie, N., Takada, Y., Shimoda, K., Miyamoto, T., Nishiwaki, T., Suda, T., & Matsuo, K.(2006). Bidirectional ephrinB2-EphB4 signaling controls bone homeostasis. *Cell Metab,* 4,111–121.

How Do Bone Cells Secrete Proteins?

Haibo Zhao, Yuji Ito, Jean Chappel, Norma Andrews, F. Patrick Ross, and Steven L. Teitelbaum

Abstract The ruffled border is the most specific marker of the active osteoclast (OC) as it forms only when the cell is resorbing bone. We provide evidence that this complex cytoskeletal structure reflects insertion of lysosomal vesicles into the bone-apposed plasma membrane under the aegis of the Ca-sensing, exocytic protein, synaptotagmin VII (SytVII). In the manner, SytVII permits transport of matrix-degrading molecules into the resorptive microenvironment. SytVII also regulates secretion of bone matrix molecules by osteoblasts. Thus, SytVII-deficient mice experience suppressed bone resorption and formation with the latter deficiency predominant thereby yielding osteoporosis characterized by attenuated remodeling.

Keywords Osteoclasts · Exocytosis · Ruffled border

Osteoclasts (OCs), which are the exclusive bone resorbing cells, degrade skeletal matrix by forming an intimate relationship with the bone surface. Thus, when OCs attach to bone, they produce an actin-rich sealing zone representing a gasket-like structure, which isolates the resorptive milieu from the general extracellular space. This "resorptive microenvironment" contains a ruffled border, the unique bone-degrading organelle of the OC, which consists of a complex, villous-like organization of the plasma membrane. This structure appears only in resorbing cells and is the product of signals derived from the bone matrix. These signals polarize as yet undefined acidified vesicles containing the OC vacuolar H^+ATPase towards the bone-apposed plasma membrane, into which they insert, thereby increasing its complexity. The ruffled border is thus the most definitive marker of the resorbing osteoclast. Its formation reflects transport of protons and chloride, which mobilize the mineral phase of bone and cathepsin K, an acidic protease degrading its organic components, into the resorptive microenvironment [1, 2]. Clearly, therefore, insertion of some type of acidifying vesicle(s) into the bone-apposed plasma membrane forms the ruffled border and promotes bone degradation [3]. Given a similar

S.L. Teitelbaum (✉)
Pathology and Immunology, Washington University School of Medicine, St. Louis, Missouri
e-mail: teitelbs@wustl.edu

Y. Choi (ed.), *Osteoimmunology*, Advances in Experimental Medicine and Biology 658, DOI 10.1007/978-1-4419-1050-9_11,

paradigm in other cells, vesicular secretion represents a likely mechanism by which OCs localize matrix-degrading molecules into the resorptive microenvironment [4]. However, the precise nature of ruffled border-forming vesicles and the mechanism by which they incorporate into the plasma membrane, are unknown.

Vesicular secretion entails attachment and incorporation of cargo-containing bodies into the plasma membrane, yielding release of their contents into the extracellular space. Binding of vesicles to the plasma membrane, in other cells, is under the aegis of SNAREs [5]. These include the VAMP, Syntaxin, SNAP, and Munc proteins. The synaptotagmins, other exocytic proteins, are unique, as they insert into both the vesicle and the plasma membranes. Synaptotagmins are calcium sensors and their interaction with SNARE proteins and phosphotidylinositols is dependent upon ambient calcium [6].

Synaptotagmin VII (Syt VII) is broadly expressed and mediates vesicle insertion in various cells [7]. We therefore asked if this synaptotagmin isoform mediates ruffled membrane genesis in the OC. To determine if SytVII is essential for secretion of bone-degrading products into the resorptive microenvironment, we visualized actin rings and cathepsin K in WT and SytVII-deficient OCs by immunoconfocal microscopy [8]. We find that actin organization is not perturbed in the absence of SytVII, as mutant cells form well-demarcated rings. However, WT but not SytVII–/– OCs secrete an abundance of cathepsin K into the actin ring-enclosed resorptive microenvironment. Whereas approximately 80% of WT OCs evidence targeted cathepsin K secretion, this number is reduced approximately three-fold in OCs lacking SytVII. In contrast to its dramatic effects on formation of the resorptive microenvironment, SytVII deficiency does not impact osteoclastogenesis as equal numbers of spread TRAP-expressing polykaryons form when WT and mutant bone marrow macrophages (BMMs) are cultured with RANK ligand (RANKL) and M-CSF. In keeping with this observation, SytVII–/– and WT cells, in osteoclastogenic conditions, express equal amounts of the OC-differentiation markers, cathepsin K, MMP-9, the β3 integrin subunit and c-Src. Thus, SytVII regulates secretion of resorptive molecules into the bone-degraded microenvironment but not in OC formation. We also find that similar to its effect in other cells, SytVII interacts with SNARE proteins, including syntaxin IV and TI-VAMP as determined by reciprocal immunoprecipitation and immunoblot.

These data indicate that SytVII participates in ruffled membrane formation in the OC. This being the case, one would expect the synaptotagmin to be located in the region of the resorptive organelle. To determine if this is so, we transduced SytVII-GFP or control vector into WT OCs. As expected, SytVII localizes within the actin ring of these cells in conjunction with cathepsin K indicating that the vesicle/membrane fusion protein is expressed at the ruffled border. Additionally, mice deficient in SytVII generate abnormal OC ruffled membranes, in vivo.

Having established that SytVII regulates ruffled border formation via vesicle insertion into the plasma membrane, we turned to the nature of the vesicles in question.

SytVII regulates calcium-dependent exocytosis of lysosomes, in fibroblasts, and delivery of these vesicular structures to phagosomes in macrophages. We therefore hypothesized that the plasma membrane-inserting vesicles in OCs are lysosome-derived [9]. This hypothesis is in keeping with the previous observation that the OC ruffled membrane contains proteins also expressed in lysosomes [10]. To address this issue, we once again expressed SytVII-GFP in WT OCs and noted that it co-localized with the lysosomal marker protein, LAMP2. To biochemically confirm the morphological observations that SytVII-regulated vesicles, which form the ruffled membrane, are lysosome-derived, we homogenized mature OCs and placed the post-nuclear supernatant in an iodixanol gradient. The mixture was centrifuged and 0.2 ml fractions collected from the bottom of the tube. The density of these fractions was measured by a refractometer. We find that, as expected, active cathepsin K co-localizes with LAMP2 documenting the enzyme is within lysosomes. Interestingly, however, the same fraction contains SytVII and additional exocytic proteins including RAB7, TI-VAMP, Sec6 and Sec8. Thus, we have documented, both morphologically and biochemically, that vesicles which form the ruffled membrane under the aegis of SytVII are lysosome-derived. As these data, in toto, indicate that SytVII is necessary for configuration of the OC ruffled membrane and thus, targeted secretion of resorptive molecules, we predicted that OCs lacking SytVII should be dysfunctional bone resorbers. We find such is the case, in vitro, as documented by resorptive pit formation and medium CTX concentration. Significantly, the same obtains, in vivo, as evidenced by serum CTX. OC number is unaltered in SytVII–/– mice, indicating that the resorptive defect does not involve arrest of osteoclastogenesis but is a manifestation of OC dysfunction. Confirming this conclusion, expression of the WT construct in SytVII–/– OCs completely rescues their bone resorptive capacity.

Inhibited bone resorption typically yields enhanced bone mass and we were therefore surprised that SytVII–/– mice are osteoporotic as determined by μCT and histomorphometry. This observation could only be explained by bone formation being compromised more than resorption. Suggesting SytVII also participates in osteoblast (OB) function, it associates with the bone matrix proteins, osteopontin and osteocalcin, and is enriched in mineralized nodules.

In keeping with the conclusion that SytVII deficiency arrests osteogenesis, calvarial OBs derived from knockout mice generate less bone nodules in culture than do their WT counterparts. However, this defect in bone formation does not reflect differentiation as WT and SytVII-deficient OBs equally express osteocalcin and alkaline phosphatase with time in culture. These data suggest that SytVII–/– OBs mature normally but are ineffective producers of bone.

Bone formation, like resorption, is a secretory process in which matrix proteins are transported extracellularly. The fact that, as in OCs, OB-residing SytVII, co-localizes with Sec6 and Sec8, as well as type I collagen, osteocalcin, and osteopontin when analyzed by ioxydol gradient, suggests SytVII also governs protein exocytosis in the context of bone formation.

Given that SytVII regulates vesicular transport in other cells, including OCs, a reasonable hypothesis would hold that the bone forming deficiency of mutant OBs reflects blunted matrix protein secretion. In keeping with this posture, extracellular accumulation of both α chains of collagen type I by SytVII–/– OBs is reduced while their pro- α counterparts accumulate intracellularly. Establishing physiological relevance to these observations bone formation is compromised in SytVII-deficient mice, as evidenced by dynamic histomorphometry and circulating osteocalcin. Vesicular secretion is therefore a major component of both bone formation and resorption and is regulated by SytVII (Fig. 1). This study underscores the importance of cell-selective modulation of exocytic proteins if they are to serve as therapeutic targets.

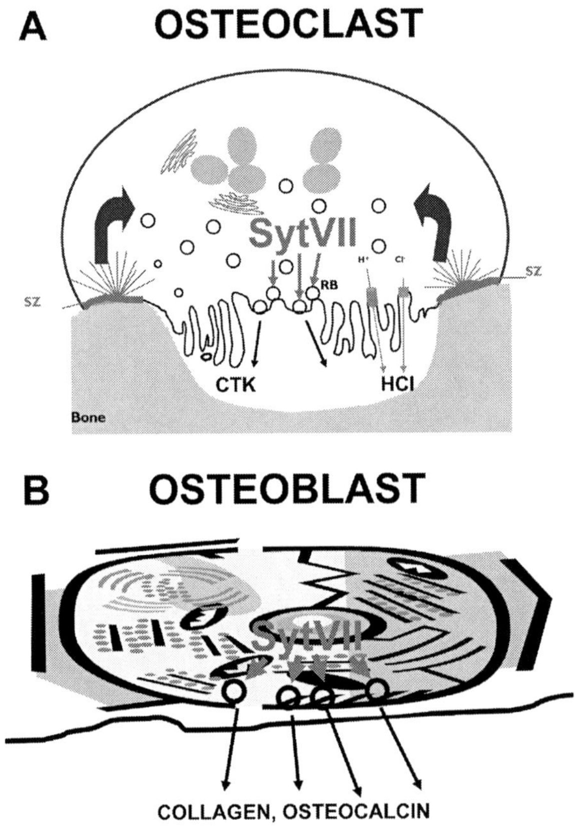

Fig. 1 Model of SytVII-regulated OC and OB function: (**A**) Upon attachment to bone, OCs form a sealing zone (SZ) (actin ring), which isolates the resorptive microenvironment from the general extracellular space. Bone derived signals (*blue arrows*) polarize lysosome-derived vesicles, containing cathepsin K, the vacuolar H^+ATPase and presumably, the Cl^- channel, to the bone-apposed plasma membrane into which they insert under the aegis of SytVII and SNARE proteins. Insertion of the vesicles markedly enhances the complexity of the targeted plasma membrane domain eventuating in formation of the ruffled border and permits transport of cathepsin K and HCl into the resorptive microenvironment; (**B**) Responding to an as yet undefined stimulus, bone matrix-containing vesicles, presumably derived from the Golgi complex, migrate to the bone-apposed plasma membrane of OBs into which they insert under the aegis of SytVII. The membrane-residing vesicles deliver their cargo resulting in bone synthesis. With time, non-directed secretion occurs resulting in incorporation of OBs into matrix as osteocytes

References

1. Andrews, N.W., & Chakrabarti, S. (2005). There's more to life than neurotransmission: the regulation of exocytosis by synaptotagmin VII. *Trends Cell Biol,* **15:**626–631.
2. Baron, R., Neff, L., Brown, W., Courtoy, P.J., Louvard, D., & Farquhar, M. (1988). Polarized secretion of lysosomal enzymes: co-distribution of cation-independent mannose-6-phosphate receptors and lysosomal enzymes along the osteoclast exocytic pathway. *J Cell Biol,* **106:**1863–1872.
3. Chieregatti, E., & Meldolesi, J. (2005). Regulated exocytosis: new organelles for non-secretory purposes. *Nature Rev,* **6:**181–187.
4. Jahn, R., & Scheller, R.H.(2006). SNAREs – engines for membrane fusion. *Nature Rev,* **7:**631–643.
5. Martinez, I., Chakrabarti, S., Hellevik, T., Morehead, J., Fowler, K., & Andrews, N.W. (2000). Synaptotagmin VII Regulates Ca2+-dependent Exocytosis of Lysosomes in Fibroblasts. *J Cell Biol,* **148:**1141–1150.
6. Rizo, J., Chen, X., & Arac, D. (2006). Unraveling the mechanisms of synaptotagmin and SNARE function in neurotransmitter release. *Trends Cell Biol,* **16:**339–350.
7. Teitelbaum, S.L. (2007). Osteoclasts: what do they do and how do they do it? *Am J Pathol,* **170:**427–435.
8. Teitelbaum, S.L., & Ross, F.P. (2003). Genetic regulation of osteoclast development and function. *Nat Rev Genet,* **4:**638–649.
9. Vaananen, H.K., Zhao, H., Mulari, M., & Halleen, J.M. (2000). The cell biology of osteoclast function. *J Cell Sci,* **113:**377–381.
10. Zhao, H., Ito, Y., Chappel, J., Andrews, N.W., Teitelbaum, S.L., & Ross, F.P. (2008). Synaptotagmin VII regulates bone remodeling by modulating osteoclast and osteoblast secretion. *Dev Cell,* **14:**914–925.

Regulation of Osteoclast Apoptosis by Bcl-2 Family Protein Bim and Caspase-3

Sakae Tanaka, Hidetoshi Wakeyama, Toru Akiyama, Katsuhiko Takahashi, Hitoshi Amano, Keiichi I. Nakayama, and Kozo Nakamura

Abstract Apoptosis of osteoclasts is regulated by the Bcl-2 family protein Bim. Bim is degraded in the course of osteoclast apoptosis, which is regulated by Caspase-3. Osteoclasts generated from *caspase-3 –/–* mice exhibited a shorter life span and a higher bone-resorbing activity than those generated from normal littermates. These results suggest the important role of Caspase-3-Bim axis in regulating both apoptosis and activation of osteoclasts.

Keywords Osteoclast · Apoptosis · Bim · Caspase-3

1 Introduction

Osteoclasts (OCs) are multinucleated giant cells primarily responsible for bone resorption; they rapidly die due to apoptosis in the absence of trophic factors such as macrophage colony-stimulating factor (M-CSF) or receptor activator of nuclear factor κB ligand (RANKL) [17, 18]. Apoptosis is genetically programmed cell death to remove the unwanted cells from physic status [6]. The abnormalities of apoptosis regulation induce various sicknesses such as cancer, autoimmune disease, and degenerative disorders [22]. Recent studies have revealed that the apoptosis of osteoclasts is strictly regulated and plays an important roles in maintaining the skeletal integrity [16].

There are two types of apoptosis pathways; the death receptor pathway and the mitochondrial pathway [13]. Bim is a pro-apoptotic BH (Bcl-2 homology) 3-only member of Bcl-2 family and induces the mitochondrial apoptosis pathway through cytochrome c release from mitochondria [5, 12]. Released cytochrome *c* interacts with Apaf-1 and Caspase-9 to form the apoptosome. Caspase-9 in the apoptosome activates effector Caspases (Caspase-3 and Caspase-7) that lead to apoptosis [15]. Bim is expressed in hematopoietic, epithelial, neuronal, and germ cells; Strasser and

S. Tanaka (✉)
Department of Orthopaedic Surgery, Faculty of Medicine, The University of Tokyo, Tokyo, Japan
e-mail: tanakas-ort@h.u-tokyo.ac.jp

Y. Choi (ed.), *Osteoimmunology*, Advances in Experimental Medicine and Biology 658, DOI 10.1007/978-1-4419-1050-9_12, © Springer Science+Business Media, LLC 2010

coworkers generated *bim*-deficient mice and demonstrated that Bim is essential for apoptosis of T lymphocytes, B lymphocytes, myeloid cells, neurons, and osteoclasts [1, 2, 3, 14]. Accumulating evidence has revealed that the expression of Bim is regulated at both transcriptional and post-translational levels. Bim is known to be regulated at the transcriptional level in hematopoietic progenitors and neurons, and the forkhead-like transcription factor FOXO3A (forkhead box O3A) was reported to be involved in the transcriptional regulation of Bim in several types of cells [4]. The other important regulation of Bim is the post-translational regulation that includes phosphorylation and ubiquitination. We and other groups previously reported that Bim is regulated via the ubiquitin-proteasome degradation process [1, 7, 8]. We found that Bim expression was markedly upregulated in the course of osteoclast apoptosis without changing its transcriptional level and was downregulated by M-CSF treatment. M-CSF maintained the protein level of Bim at low levels by inducing its ubiquitination, which was at least partly mediated by an E3 ubiquitin ligase c-Cbl although there is a controversy about the role of c-Cbl in Bim degradation [21].

2 Postranslational Regulation of Bim in MEF/Bcl-2 Cells

To investigate the molecular mechanisms underlying the degradation of Bim in detail, we analyzed the protein dynamics of Bim in mouse embryonic fibroblasts (MEF) [19]. Since we found it difficult to overexpress BimEL in MEF cells, probably due to its strong proapoptotic activity, we generated MEF cells in which the expression level of Bcl-2 can be regulated by the Tet-off system (MEF/Bcl-2 cells). BimEL can be stably overexpressed in the cells without causing cell death. Doxycycline treatment reduced mRNA and protein levels of Bcl-2 in a time-dependent manner. After 12 h of doxycycline treatment, activation of Caspase-3 was observed, which was associated with the decrease in Bim protein levels (Fig. 1A) without affecting its mRNA level (data not shown). The protein levels of

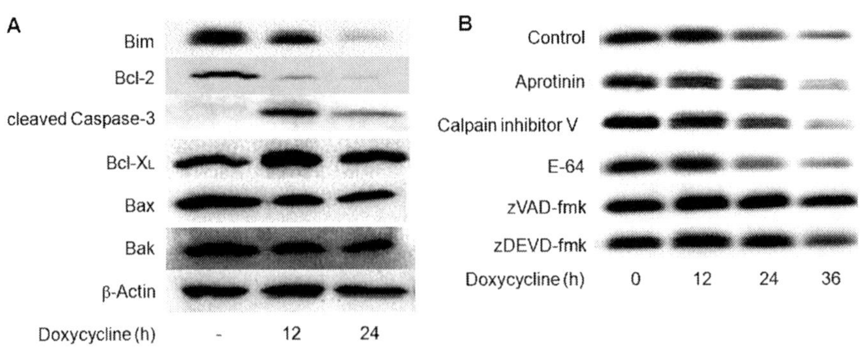

Fig. 1 Regulation of Bim by Caspase-3 in MEF/Bcl-2 cells. (**A**) Doxycycline treatment induced downregulation of Bcl-2, which led to degradation of Bim in MEF Bcl-2 Tet-off cells overexpressing Bim. (**B**) Bim degradation in MEF/Bcl-2 cells overexpressing Bim was suppressed by a pan Caspase inhibitor zVAD-fmk or a Caspase-3-specific inhibitor zDEVD-fmk, but not by aprotinin, calpain inhibitor V or E-64

other Bcl-2 family members did not appear to alter during this period (Fig. 1A). The degradation of Bim was not affected by aprotinin, calpain inhibitor V or cathepsin inhibitor E64 but was inhibited by a broad-spectrum Caspase inhibitor zVAD-fmk or a Caspase-3-specific inhibitor zDEVD-fmk (Fig. 1B) [19].

3 Bim Expression Is Downredulated by Caspase-3 in Osteoclasts

We next investigated whether similar regulation of Bim is observed in osteoclasts (OCs) [20]. OCs generated from mouse bone marrow cells in the presence of recombinant human M-CSF (10 ng/ml) and soluble RANKL (100 ng/ml) underwent cell death within 48 h after removal of these cytokines [10]. The protein level of Bim increased after 12 h of the cytokine removal and reduced again after 24 h while the expression of other Bcl-2 family members did not appear to change (Fig. 2A). The reduction of Bim levels was preceded by the activation of Caspase-3 and zVAD-fmk and Lactacystin; however, no other proteinase inhibitors maintained Bim at high levels after 24 h of the cytokine removal (data not shown). To further confirm the role of Caspase-3 on Bim degradation in OCs, we generated OCs from *caspase-3*-deficient (*caspase-3 –/–*) mouse bone marrow cells. The degradation of Bim was much reduced in *caspase-3 –/–* OCs, and a high level of Bim was maintained 24 h after the cytokine withdrawal (Fig. 2B). These results suggest that Caspase-3 is critically involved in the degradation of Bim in OCs as well.

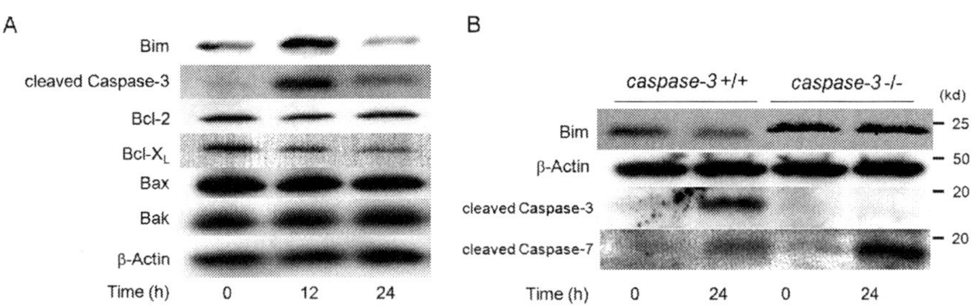

Fig. 2 Involvement of Caspase-3 in Bim degradation in osteoclasts. (**A**) OCs underwent apoptosis after 24 h of cytokine withdrawal. Bim expression levels were increased 12 h after the cytokine deprivation, and then decreased 24 h in association with Caspase-3 activation, while the expression of other Bcl-2 family members did not change. (**B**) Bim degradation was suppressed in *caspase-3*-deficient OCs. OCs were generated from bone marrow cells of *caspase-3 –/–* mice or their normal littermates. Bim degradation was not observed 24 h after the cytokine removal in *caspase-3 –/–* OCs. Caspase-7 activation was observed in *caspase-3 –/–* OCs at the comparable level as control OCs. Adapted from J Bone Miner Res 2007: 22; 1631–1639 with permission of the American Society for Bone and Mineral Research.

4 Caspase-3 Regulates Survival and Activation of Osteoclasts

We next examined the effect of *caspase-3* deficiency on the survival and the activation of OCs. *Caspase-3 –/–* mice (on a C57BL/6 genetic background) were generated as previously reported [11]. There was no significant difference in OC differentiation between bone marrow cells from *caspase-3* +/+ and *caspase-3* –/– mice (data not shown). In contrast, the survival of OCs generated from *caspase-3* –/– bone marrow cells exhibited a shorter life span and a higher bone-resorbing activity than the OCs generated from normal littermates (Fig. 3). These results were consistent with the in vivo observations which showed that the number of OCs is reduced in the bone tissues of *caspase-3* –/– mice which were recently reported by other groups [9].

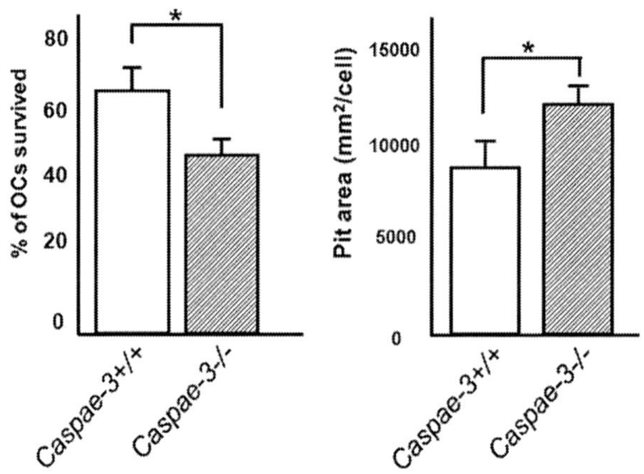

Fig. 3 *Caspases-3* –/– OCs exhibited a shorter life span and a higher bone-resorbing activity than *caspases-3* +/+ OCs. OCs were generated from bone marrow cells of *caspases-3* –/– mice or their normal littermates (C57BL/6 genetic background), and subjected to the survival assay and the bone resorption assay. Survival rate after 16 h incubation (*left*) and pit area per cell (μm^2/cell) formed on dentine slices (*right*) are shown. *Caspase-3* –/– OCs exhibited a significantly lower survival rate and a higher bone-resorbing activity than control OCs. Error bars represent standard deviation. ∗significantly different, $p<0.01$. Adapted from J Bone Miner Res 2007: 22; 1631–1639 with permission of the American Society for Bone and Mineral Research.

5 Conclusion

In this study, we demonstrated the involvement of Caspase-3 in the posttranslational regulation of Bim. Bim activates Caspase cascades by inducing cytochrome *c* release from the mitochondria and induces apoptosis of the cells. In turn, Caspase-3 regulates Bim degradation, thus creating a negative feedback loop (Fig. 4); however, the molecular machinery how Caspase-3 activation induces the degradation of Bim still remains unclear. Since osteoclasts are primary cells for bone resorption, the regulation of their apoptosis is critical for maintaining skeletal homeostasis and can be a potential therapeutic target. It should be noted that Bim regulates bone-resorbing activity as well as apoptosis of OCs. Further study is required to elucidate the detailed mechanisms of Bim regulation and its importance in osteoclast biology.

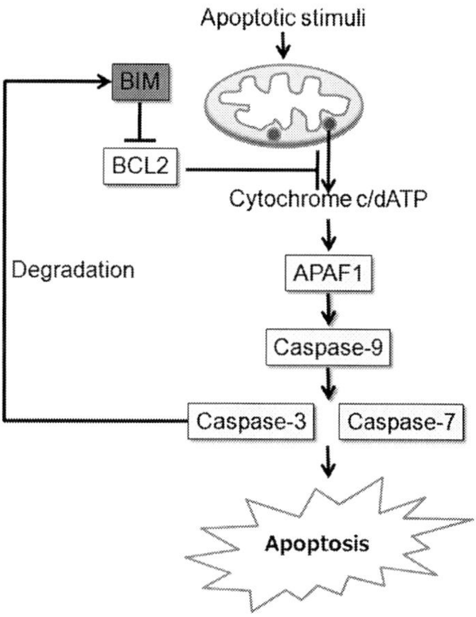

Fig. 4 Possible role of Caspase-3 in Bim degradation. Apoptotic stimuli induce cytochrome *c* release from mitochondria, which is considered to be caused by the inhibition of Bcl-2 by Bim, and subsequently causes Caspase-3 and Caspase-7 activation. Activated Caspase-3 leads to the degradation of Bim, thus creating a negative feedback loop

Acknowledgments This work was supported in part by Grants-in-Aid from the Ministry of Education, Culture, Sports, Science, and Technology of Japan and the Health Science research grants from the Ministry of Health, Labor, and Welfare of Japan to S.T.

References

1. Akiyama, T., Bouillet, P., Miyazaki, T., Kadono, Y., Chikuda, H., Chung, U. I. et al. (2003). Regulation of osteoclast apoptosis by ubiquitylation of proapoptotic BH3-only Bcl-2 family member Bim. *Embo J* 22:6653–6664.
2. Bouillet, P., Metcalf, D., Huang, D. C., Tarlinton, D. M., Kay, T. W., Kontgen, F. et al. (1999). Proapoptotic Bcl-2 relative Bim required for certain apoptotic responses, leukocyte homeostasis, and to preclude autoimmunity. *Science* 286:1735–1738.
3. Bouillet, P., Purton, J. F., Godfrey, D. I., Zhang, L. C., Coultas, L., Puthalakath, H. et al. (2002). -only Bcl-2 family member Bim is required for apoptosis of autoreactive thymocytes. *Nature* 415:922–926.
4. Gilley, J., Coffer, P. J., & Ham, J. (2003). FOXO transcription factors directly activate bim gene expression and promote apoptosis in sympathetic neurons. *J Cell Biol* 162:613–622.
5. Huang, D. C., & Strasser, A. (2000). BH3-Only proteins-essential initiators of apoptotic cell death. *Cell* 103:839–842.
6. Kerr, J. F., Wyllie, A. H., & Currie, A. R. (1972). Apoptosis: a basic biological phenomenon with wide-ranging implications in tissue kinetics. *Br J Cancer* 26:239–257.
7. Ley, R., Balmanno, K., Hadfield, K., Weston, C. R., & Cook, S. J. (2003). Activation of the ERK1/2 signalling pathway promotes phosphorylation and proteasome-dependent degradation of the BH3-only protein Bim. *J Biol Chem* 278:18811–18816.
8. Meller, R., Cameron, J. A., Torrey, D. J., Clayton, C. E., Ordonez, A. N., Henshall, D. C. et al. (2006). Rapid degradation of Bim by the ubiquitin-proteasome pathway mediates short-term ischemic tolerance in cultured neurons. *J Biol Chem* 281:7429–7436.

9. Miura, M., Chen, X. D., Allen, M. R., Bi, Y., Gronthos, S., Seo, B. M. et al. (2004). A crucial role of caspase-3 in osteogenic differentiation of bone marrow stromal stem cells. *J Clin Invest* 114:1704–1713.
10. Miyazaki, T., Katagiri, H., Kanegae, Y., Takayanagi, H., Sawada, Y., Yamamoto, A. et al (2000). Reciprocal role of ERK and NF-kappaB pathways in survival and activation of osteoclasts. *J Cell Biol* 148:333–342.
11. Morishita, H., Makishima, T., Kaneko, C., Lee, Y. S., Segil, N., Takahashi, K. et al. (2001). Deafness due to degeneration of cochlear neurons in caspase-3-deficient mice. *Biochem Biophys Res Commun* 284:142–149.
12. O'Connor, L., Strasser, A., O'Reilly, L. A., Hausmann, G., Admas, J. M., Cory, S. et al. (1998). Bim: a novel member of the Bcl-2 family that promotes apoptosis. *Embo J* 17: 384–395.
13. Opferman, J. T., & Korsmeyer, S. J. (2003). Apoptosis in the development and maintenance of the immune system. *Nat Immunol* 4:410–415.
14. Putcha, G. V., Moulder, K. L., Golden, J. P., Bouillet, P., Adams, J. A., Strasser, A. et al. (2001). Induction of BIM, a proapoptotic BH3-only BCL-2 family member, is critical for neuronal apoptosis. *Neuron* 29:615–628.
15. Siegel, R. M. (2006). Caspases at the crossroads of immune-cell life and death. *Nat Rev Immunol* 6:308–317.
16. Tanaka, S. (2007). Signaling axis in osteoclast biology and therapeutic targeting in the RANKL/RANK/OPG system. *Am J Nephrol* 27:466–478.
17. Tanaka, S., Nakamura, I., Inoue, J., Oda, H., & Nakamura, K. (2003). Signal transduction pathways regulating osteoclast differentiation and function. *J Bone Miner Metab* 21:123–133.
18. Tanaka, S., Nakamura, K., Takahasi, N., & Suda, T. (2005). Role of RANKL in physiological and pathological bone resorption and therapeutics targeting the RANKL-RANK signaling system. *Immunol Rev* 208:30–49.
19. Wakeyama, H., Akiyama, T., Kadono, Y., Nakamura, M., Oshima, Y., Nakamura, K. et al. (2007a). Posttranslational regulation of Bim by caspase-3. *Ann N Y Acad Sci* 1116:271–280.
20. Wakeyama, H., Akiyama, T., Takahashi, K., Amano, H., Kadono, Y., Nakamura, M. et al. (2007b). Negative feedback loop in the Bim-caspase-3 axis regulating apoptosis and activity of osteoclasts. *J Bone Miner Res* 22:1631–1639.
21. Wiggins, C. M., Band, H., & Cook, S. J. (2007). c-Cbl is not required for ERK1/2-dependent degradation of BimEL. *Cell Signal* 19:2605–2611.
22. Wyllie, A. H., Kerr, J. F., & Currie, A. R. (1980). Cell death: the significance of apoptosis. *Int Rev Cytol* 68:251–306.

Regulation of Bone Formation and Immune Cell Development by Schnurri Proteins

Dallas C. Jones and Laurie H. Glimcher

Abstract Although identified over a decade ago, the function and physiological significance of the mammalian Schnurri protein family remained largely unknown. However, the recent generation and characterization of mice bearing null mutations in the individual Schnurri genes has led to the discovery of unexpected yet central roles for these large zinc-finger proteins in several biological processes. Here, we review findings of these studies and discuss the importance of the Schnurri protein family in regulating both the immune and skeletal systems.

Keywords Schnurri · Skeletal system · Immune system · Osteoblasts · T lymphocytes

1 Introduction

Skeletal patterning begins during embryogenesis when mesenchymal stem cells condensate into elements that prefigure the future bone [16]. Formation of the embryologic skeleton then proceeds through two distinct processes: endochondral and intramembranous ossification [11]. The latter process occurs when mesenchymal condensates differentiate directly into osteoblasts that then synthesize the extracellular matrix (ECM) constituents that form the flat bones in skull and medial clavicle. The remainder of the axial and appendicular skeleton is formed through endochondral ossification in which cartilage is replaced by bone. Chondrocytes that are located in the growth plate of the developing bones synthesize the ECM components required to form the cartilaginous template. As the chondrocytes reach the terminal stage of differentiation, they produce additional factors that promote neovasuclarization of the cartilaginous scaffold that allows for an influx of cells required to complete the ossification process [15].

D.C. Jones (✉)
Department of Infectious Disease and Immunology, Harvard School of Public Health, Boston, MA USA
e-mail: djones@hsph.harvard.edu

Y. Choi (ed.), *Osteoimmunology*, Advances in Experimental Medicine and Biology 658, DOI 10.1007/978-1-4419-1050-9_13, © Springer Science+Business Media, LLC 2010

Cells of hematopoietic origin are among those that colonize the newly vascularized cartilage, thus presenting the initial interface between the skeletal and immune systems. The interaction between these two organ systems has evolved into one that is of mutual benefit. The further development and remodeling of the skeleton system requires cells of hematopoietic origin, specifically osteoclasts, which are essential for bone resorption [16]. Reciprocally, the skeletal system provides a vital niche that supports hematopoiesis [12].

As the interaction between the skeletal system and immune system has been interrogated, common gene products have been identified that regulate both organ systems [23]. Herein, we review the emerging role for the Schnurri protein family in regulating bone formation and immune cell function.

2 Schnurri Protein Family

Proteins in the Schnurri family belong to a unique class of zinc-finger proteins that are characterized by the presence of a separated pair of C_2H_2 zinc-fingers (Fig. 1) [8]. Proteins containing this motif often function as scaffolding proteins that facilitate the formation of multicomponent complexes on DNA. Indeed, the Drosophila Shn protein mediates a similar scaffolding function in regulating gene expression downstream of decapentaplegic (Dpp), a member of the TGFβ superfamily, during Drosophila development [2, 6]. Ligation of Dpp to its receptors initiates a signal cascade that results in Med, the *Drosophila* Co-Smad homologue, partnering with Mad, the *Drosophila* R-Smad homologue [1]. The Mad/Med complex translocates to the nucleus where it can interact with Shn. The presence of Shn in this multimeric complex is essential for Dpp-mediated repression of specific genes as it recruits the necessary transcriptional co-repressors to the Mad/Med complex [20]. The elucidation of Shn's function during Drosophila development has been of great value in understanding the function of this protein family in higher organisms.

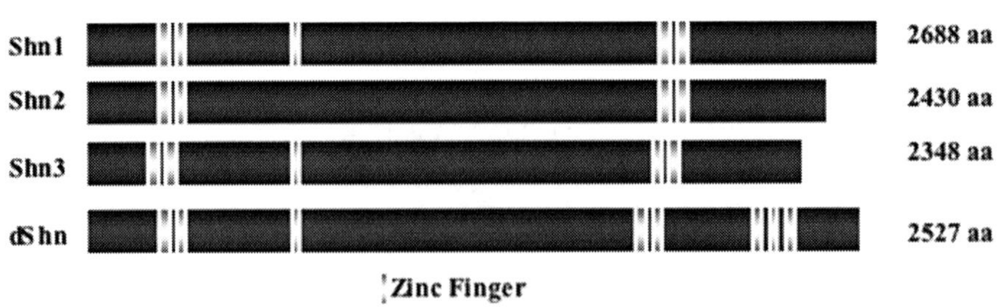

Fig. 1 Linear Schematic of Schnurri Proteins. Shn proteins are characterized by the presence of a pair of C_2H_2 zinc-fingers in both the N-terminus and C-terminus. In addition to these paired zinc fingers, the mammalian Shn1 and Shn3 proteins contain an additional fifth zinc finger in the middle portion of the protein. The drosophila Shn protein contains a middle zinc-finger as well as three additional C-terminal zinc-fingers

Each of the mammalian Shn proteins (Shn1, Shn2, and Shn3) were independently identified as large DNA-binding proteins in lymphocytes [3, 5, 17, 24]. These early findings placed an initial interest in understanding how these proteins function within the immune system. However, subsequent studies revealed broad expression patterns for each Shn gene that resulted in overlapping expression in multiple tissue and cell types, suggesting that these proteins may be involved in multiple biological processes. The physiological relevance of these proteins has only recently become evident with the generation of mice bearing null-mutations in the individual Shn genes. As we review below, analysis of these genetically engineered mice has been essential in uncovering a central role for the Shn protein family in both skeletal and immune systems.

3 Regulation of Immune Cell Function by Schnurri Proteins

The ability of thymocytes to interpret signals emanating from the T-cell receptor (TCR) following engagement with antigen loaded major-histocompatibility complexes are essential for T cells to properly develop [4]. Analysis of mice lacking Shn2 revealed a profound defect in thymopoiesis suggesting that Shn2 assists in interpreting these signals in developing T cells [22]. Histological analysis of thymi isolated from Shn2–/– mice revealed an abnormal thymic architecture characterized by a large cortex and a small medulla. Further characterization of the individual cell populations within the Shn2–/– thymi showed similar numbers of CD4/CD8 double negative (DN) T cells and CD4/CD8 double positive (DP) T cells when compared to WT thymi. However, there was a significant reduction in CD4+ single positive and CD8+ single positive T cells in the Shn2–/– thymi. To fully demonstrate that Shn2 was regulating positive selection through its function in the thymocyte population and not in the stromal population, bone marrow chimeras were generated in which bone marrow from Shn2–/– mice was transferred into lethally irradiated WT hosts. Conversely, lethally irradiated Shn2–/– hosts received WT bone marrow. Normal levels of CD4-SP and CD8-SP levels were observed in Shn2–/– hosts that received the WT bone marrow, whereas transfer of Shn2–/– bone marrow into WT recipients recapitulated the defect in thymopoiesis that was originally observed in the Shn2–/– mice. Collectively, these results suggested that Shn2 expression within developing thymocytes is essential for positive selection of T cells to occur.

Although Shn2–/– mice exhibit a defect in thymopoiesis, a moderate number of CD4+ T cells and CD8+ T cells are observed in secondary lymphoid organs [22, 13]. Subsequent analysis of the peripheral CD4+ T cell populations in the Shn2–/– mice uncovered an enhanced skewing of the cells towards the Th2 subset when compared to CD4+ T cells isolated from WT mice [13]. Molecular analysis of CD4+ T cells isolated from Shn2–/– hosts revealed a hyper-activation of NF-κB in response to TCR stimulation. The authors suggested that the enhanced NF-κB activation and transcriptional activity in stimulated Shn2–/– CD4+ T cells may be driving the elevated expression of GATA-3, a transcription factor that is required for Th2 skewing, that is observed in Shn2–/– CD4+ T cells. It was further demonstrated that Shn2 regulates the transcriptional activity of NF-κB through competitive binding of κB sites. Therefore, NF-κB had increased access to the κB binding sites in certain genes in the absence of Shn2 that not only control the generation of Th2 T cells but also the generation of memory T cells [14].

In striking similarity to Shn2, it has been demonstrated that Shn3 is highly expressed in the thymus and regulates gene expression through antagonism of NF-κB [18, 7]. However, analysis of Shn3–/– mice by our group has only demonstrated a unique role for Shn3 in peripheral lymphocyte populations, as thymopoiesis is unaffected in these mice. Isolation of CD4+ T cells from Shn3–/– mice produced less IL-2 but normal levels of IFN-g in response to anti-CD3/anti-CD28 stimulation as compared to WT CD4+ T cells [19]. Further in vitro experiments revealed that Shn3 regulates IL-2 production by T cells at the level of gene transcription. Shn3 was demonstrated to regulate IL-2 gene transcription through it association with c-Jun that augments AP-1 transcriptional activity when bound to the IL-2 promoter. These studies were the first to demonstrate an *in vivo* role for Shn3 in the immune system and further emphasized the importance that the Shn2–/– mice and Shn3–/– mice have been to uncovering previously unknown roles for the Shn proteins in the immune system.

4 Schnurri Proteins and Skeletal Biology

During our initial studies exploring the function of Shn3 in the immune system, we made the serendipitous observation that the Shn3–/– mice have a profound increase in bone mass [10]. The osteosclerotic phenotype observed in Shn3–/– mice arises through augmented osteoblast activity and is characterized by greatly increased rates of bone formation that results in an age-associated progression of the phenotype. However, Shn3–/– mice exhibit normal skeletal morphology, as the onset of the osteosclerotic phenotype is postnatal. Our initial in vitro studies identified that Shn3 regulates osteoblast activity by controlling protein levels of Runx2, a central regulator of osteoblast biology. Shn3 associates with and augments the activity of the E3 ubiquitin ligase WWP1. Thus the association of Shn3 with WWP1 in osteoblasts increases turn over of Runx2 through ubiquitin-mediated degradation. While these studies demonstrate that Shn3 is an essential regulator of postnatal skeletal remodeling, numerous questions still exist in regard to the function of this protein in regulating various aspects of skeletal biology. It will be of importance to determine if Shn3 function and/or expression is regulated by any external signaling pathways. Following the elucidation that the drosophila Shn protein regulates expression of Dpp target genes, it was theorized that mammalian Shn proteins might also function within the TGFβ/BMP signaling cascade. Given the important role of the Bmp/TGFβ family in skeletal biology, it is tempting to speculate that Shn3 may function downstream of these signaling molecules in osteoblasts.

Interestingly, validation that mammalian Shn proteins regulate the expression of BMP/TGFβ target genes arose from studies examining the role of Shn2 in adipogenesis [9]. Mice lacking Shn2 have a reduction in white adipose tissue due to a defect in adipocyte differentiation. It was demonstrated that Shn2 associates with Smad1 following Bmp2 stimulation to drive expression of PPARγ, a master regulator of adipogenesis. Given the more established role for Bmp2 in promoting bone formation, Shn2 was subsequently analyzed for its role in regulating bone formation [21]. Analysis of bone marrow stromal cultures from Shn2–/– mice revealed decreased osteoblastic potential as they gave rise to fewer mineralized nodules when compared to WT bone marrow stromal cultures. Similar to the observations made in adipocytes, the authors presented data to suggest that Shn2 expression in osteoblasts augments these cells response to Bmp2. Shn2 was also shown to be required for optimal osteoclast differentiation *in vitro*. The combination of reduced osteoblast and osteoclast function results in a low-turnover osteopenia in Shn2–/– mice. These results stand in stark contrast to the osteosclerotic phenotype observed in the Shn3–/– mice. Given the potential "scaffolding" function of the Schnurri proteins, identifying the various proteins that associate with Shn2 and Shn3 will be necessary to understand how different Shn proteins elicit distinct cellular responses in both the skeletal and immune systems.

5 Conclusions

The Schnurri family of large zinc-finger proteins are present in numerous tissues throughout the body often resulting in multiple Shn proteins being present in a given cell type. However, analysis of mice deficient for a single Schnurri gene has established that individual Schnurri proteins possess unique roles in regulating multiple physiological processes, including lymphocyte development and bone formation.

Mice with deletions in individual Shn genes may not reveal those biological processes where overlapping expression of multiple Shn genes result in functional redundancy. Therefore, analysis of mice deficient in multiple Schnurri proteins will be required to gain a comprehensive understanding of the biological process regulated by this protein family.

References

1. Affolter, M., T. Marty, M. A. Vigano, & A. Jazwinska (2001). Nuclear interpretation of Dpp signaling in Drosophila. *Embo J* **20**(13): 3298–2305.
2. Arora, K., H. Dai, S. G. Kazuko, J. Jamal, M. B. O'Connor, A. Letsou, & R. Warrior (1995). The Drosophila schnurri gene acts in the Dpp/TGF beta signaling pathway and encodes a transcription factor homologous to the human MBP family. *Cell* **81**(5): 781–790
3. Baldwin, A. S., Jr., K. P., LeClair, H. Singh, & P. A. Sharp (1990). A large protein containing zinc finger domains binds to related sequence elements in the enhancers of the class I major histocompatibility complex and kappa immunoglobulin genes.*Mol Cell Biol* **10**(4): 1406–1414.
4. Boehm, T. (2008). Thymus development and function. *Curr Opin Immunol* **20**(2): 178–184.

5. Fan, C. M., & T. Maniatis (1990). A DNA-binding protein containing two widely separated zinc finger motifs that recognize the same DNA sequence. *Genes Dev* **4**(1): 29–42.
6. Grieder, N. C., D. Nellen, R. Burke, K. Basler, & M. Affolter (1995). Schnurri is required for Drosophila Dpp signaling and encodes a zinc finger protein similar to the mammalian transcription factor PRDII-BF1. *Cell* **81**(5): 791–800.
7. Hong, J. W., C. E. Allen, & L. C. Wu (2003). Inhibition of NF-kappaB by ZAS3, a zinc-finger protein that also binds to the kappaB motif. *Proc Natl Acad Sci U S A* **100**(21): 12301–12306.
8. Iuchi, S. (2001). Three classes of C_2H_2 zinc finger proteins.*Cell Mol Life Sci* **58**(4): 625–635.
9. Jin, W., T. Takagi, S. N. Kanesashi, T. Kurahashi, T. Nomura, J. Harada, & S. Ishii (2006). Schnurri-2 controls BMP-dependent adipogenesis via interaction with Smad proteins. *Dev Cell* **10**(4): 461–471.
10. Jones, D. C., M. N. Wein, M. Oukka, J. G. Hofstaetter, M. J. Glimcher, & L. H. Glimcher (2006). Regulation of adult bone mass by the zinc finger adapter protein Schnurri-3. *Science* **312**(5777): 1223–1227.
11. Karsenty, G., & E. F. Wagner (2002). Reaching a genetic and molecular understanding of skeletal development. *Dev Cell* **2**(4): 389–406.
12. Kiel, M. J., & S. J. Morrison (2008). Uncertainty in the niches that maintain haematopoietic stem cells. *Nat Rev Immunol* **8**(4): 290–301.
13. Kimura, M. Y., H. Hosokawa, M. Yamashita, A. Hasegawa, C. Iwamura, H. Watarai, M. Taniguchi et al. (2005). Regulation of T helper type 2 cell differentiation by murine Schnurri-2. *J Exp Med* **201**(3): 397–408.
14. Kimura, M. Y., C. Iwamura, A. Suzuki, T. Miki, A. Hasegawa, K. Sugaya, M. Yamashita et al. (2007). Schnurri-2 controls memory Th1 and Th2 cell numbers in vivo. *J Immunol* **178**(8): 4926–4936.
15. Kronenberg, H. M. (2003). Developmental regulation of the growth plate. *Nature* **423**(6937): 332–336.
16. Mariani, F. V., & G. R. Martin (2003). Deciphering skeletal patterning: clues from the limb. *Nature* **423**(6937): 319–325.
17. Nomura, N., M. J. Zhao, T. Nagase, T. Maekawa, R. Ishizaki, S. Tabata, & S. Ishii (1991). HIV-EP2, a new member of the gene family encoding the human immunodeficiency virus type 1 enhancer-binding protein. Comparison with HIV-EP1/PRDII-BF1/MBP-1. *J Biol Chem* **266**(13): 8590–8594.
18. Oukka, M., S. T. Kim, G. Lugo, J. Sun, L. C. Wu, & L. H. Glimcher (2002). A mammalian homolog of Drosophila schnurri, KRC, regulates TNF receptor-driven responses and interacts with TRAF2. *Mol Cell* **9**(1): 121–131.
19. Oukka, M., M. N. Wein, & L. H. Glimcher (2004). Schnurri-3 (KRC) interacts with c-Jun to regulate the IL-2 gene in T cells. *J Exp Med* **199**(1): 15–24.
20. Pyrowolakis, G., B. Hartmann, B. Muller, K. Basler, & M. Affolter (2004). A simple molecular complex mediates widespread BMP-induced repression during Drosophila development. *Dev Cell* **7**(2): 229–240.
21. Saita, Y., T. Takagi, K. Kitahara, M. Usui, K. Miyazono, Y. Ezura, K. Nakashima et al. (2007). Lack of Schnurri-2 expression associates with reduced bone remodeling and osteopenia. *J Biol Chem* **282**(17): 12907–12915.
22. Takagi, T., J. Harada, & S. Ishii (2001). Murine Schnurri-2 is required for positive selection of thymocytes. *Nat Immunol* **2**(11): 1048–1053.
23. Wein, M. N., D. C. Jones, & L. H. Glimcher (2005). Turning down the system: counter-regulatory mechanisms in bone and adaptive immunity. *Immunol Rev* **208**: 66–79.
24. Wu, L. C., C. H. Mak, N. Dear, T. Boehm, L. Foroni, & T. H. Rabbitts (1993). Molecular cloning of a zinc finger protein which binds to the heptamer of the signal sequence for V(D)J recombination. *Nucleic Acids Res* **21**(22): 5067–5073.

Index

Y. Choi (ed.), *Osteoimmunology,* Advances in Experimental Medicine and Biology 658, DOI 10.1007/978-1-4419-1050-9, © Springer Science+Business Media, LLC 2010